PRAISE FOR
THE OIL FACTOR

"A truly impressive account of how much the oil supply problem has been underestimated and misunderstood. THE OIL FACTOR cannily guides investors faced with the prospect of an alarming, all-too-likely scenario."
—**GENE G. MARCIAL, senior writer and**
"Inside Wall Street" columnist, *BusinessWeek*

"Brazen, brilliant . . . and a little bit frightening. Provocative and prophetic, this is one of the most important books I've read in years. Those who heed its strategies will be richly rewarded."
—**JONATHAN HOENIG, portfolio manager,**
Capitalistpig Hedge Fund, LLC
and author of *Greed Is Good*

"Destined to be viewed as a classic text when the financial history of our era is written. Never let it be said that we weren't warned . . . THE OIL FACTOR is an absolutely indispensable tool in every investor's arsenal."
—**THOMAS KAPLAN, chairman, Apex Silver, and**
Oxford PhD in History

"Once again, Leeb and his wife, Donna, are ahead of the curve. A bold look at the next energy crunch . . . this book beams in on how to invest during seismic shifts."
—**STEVE KICHEN, assistant managing editor,**
Forbes **magazine**

more . . .

"A must read . . . right on! Based on solid research . . . provocative . . . a crucial addition to the inflation/deflation debate . . . [that] also presents wonderful investment opportunities."
—MICHAEL A. BERRY, PhD, formerly professor, University of Virginia, Darden School

"Thought provoking . . . Features valuable advice on how investors cannot just protect themselves but profit from coming oil scarcity. A thoroughly enjoyable read."
—ROGER S. CONRAD, editor, *Utility Forecaster*

"THE OIL FACTOR paints a bull's-eye squarely on the biggest problem facing the American economy in the early twenty-first century."
—HOWARD M. CROSBY, president, Cadence Resources Corporation

"Excellent . . . Filled with great advice from experts [and] written in an easy-to-understand language."
—BestsellersWorld.com

"Discusses a diverse strategy to stay ahead of the game during the volatile years ahead."
—*Booklist*

"This book offers a solid, concise overview of the economy and the stock trends."
—*Publishers Weekly*

"Insightful and compelling . . . easy to read. The discussion of the Oil Indicator alone is worth the price of the book."
—Suite101.com

THE Oil
FACTOR

also by Stephen Leeb and Donna Leeb

Getting in on the Ground Floor

Defying the Market

by Stephen Leeb and Roger Conrad

Market Timing for the Nineties

The Agile Investor

THE Oil FACTOR

Protect Yourself—AND PROFIT—
from the Coming Energy Crisis

STEPHEN LEEB
and DONNA LEEB

**BUSINESS
PLUS**

NEW YORK   BOSTON

Book design by Fearn Cutler deVicq

Business Plus
Hachette Book Group USA
237 Park Avenue
New York, NY 10017
Visit our Web site at www.HachetteBookGroupUSA.com

Business Plus is an imprint of Grand Central Publishing. The Business Plus name and logo is a trademark of Hachette Book Group USA, Inc.

Printed in the United States of America

Originally published in hardcover by Hachette Book Group USA
First trade edition: February 2005
First international mass market edition: September 2008

10 9 8 7 6 5 4 3 2 1

To Tim and Will

Acknowledgments

We gratefully acknowledge the hard work and contributions of Genia Turanova and Toby Crabtree in helping research stocks and funds. Equal thanks to Steve Kizer for the charts and tables.

We are thankful that we were lucky enough to link up with our proleptic editor Rick Horgan, who believed in this project from the beginning and whose astute comments and questions continually steered us in the right direction.

As always, we're grateful to Al Zuckerman, our agent at Writer's House, who kept pushing us to make the proposal better.

Finally, we want to thank everyone at Leeb Capital Management and Leeb Brokerage Services for being such hard workers as well as such great people. There are too many of you to mention everyone individually, but we truly appreciate all of you. We can't not mention Steve Fishman for handling pretty much everything; Bob Lehr for his friendship and always astute advice; Steve Perkins and Wade Black for their strong support; and Burt Dorsett for his valuable understanding of markets. Finally, thanks to Regina Fishman, Jim Rutz, and Peter Bohning.

Contents

Part II: Making Money

List of Figures

Preface

Investing is a tough business, whether you do it simply for your own account or professionally. And one of the toughest things about it is that there is so much information out there that conceivably can bear upon your decisions. To be a savvy investor, there's nothing that isn't relevant, from science to economics to international relations to politics in Washington. It all bears upon the economy, upon the financial markets, and upon prospects for individual stocks. Whew!

As we said, this makes it tough—but it also makes it wonderfully fascinating and fulfilling, because if you take it seriously, it plunges you into the whole wide world of knowledge. As part and parcel of our investment responsibilities, we devour not just daily newspapers and the basic financial publications but also a slew of science, technology, medical, and political magazines and journals, not to mention books and, now, Internet articles.

Then, of course, the task becomes one of paring down and synthesizing. Most of the information we absorb is interesting, but not all of it is equally useful. Knowing how to evaluate what you read, and how to draw together disparate threads of information and apply them to prospects for the financial markets, is the real challenge. But it's also what we really love to do.

Since we've been in this business, for over a quarter of a century, we've always been drawn to trying to see the big picture. It's like those digitally created picture puzzles that were popular a few years back. When you first look at a page, it's just a collection of colors and dots. If you peer a little longer, though, and relax your eyes in just the right way, suddenly a three-dimensional image emerges. When it comes to the investment picture, we want to avoid being overwhelmed by the dots. We want to see what lies behind them—what they add up to.

In our first book, *Getting in on the Ground Floor,* the big-picture trend we saw was declining inflation. We understood that low inflation is what makes it possible for the economy to sustain growth and that prospects of sustainable economic growth are what lead to bull markets. Thus we predicted that the market was about to enter a long-term bull phase that would carry the Dow to above 4000. We began writing the book in 1984 when the Dow was a bit above 1000, and many people thought we were nuts.

Our most recent book before this one, *Defying the Market: Profiting in the Turbulent Post-Technology Market Boom,* also was a big-picture type of effort. We wrote it in 1998, as the tech boom was really gathering steam, and it was published in midsummer 1999. The key trend we identified was that—contrary to what almost everyone else thought—technological progress was slowing down. As a result, we warned the tech boom wasn't sustainable. Half a year later, the tech boom turned into a bust.

One reason why we gravitate to the big picture is that it enables us to offer the best investment advice we can. In *Defying the Market,* we used our big-picture insights to construct a model portfolio. We recommended specific stocks and specific allocations—no fudging. For the four-year period after the book came out, our recommendations rose by some 10 percent, compared to a loss of nearly 30 percent in the S&P 500. That hefty outperformance resulted from our insight into the big picture, an understanding of what was really important in the overall investment scheme of things.

This book, *The Oil Factor,* is another big-picture book, and we think it's our most important one yet. The underlying trend it deals with relates to energy, which is as essential to our economy and society as food is to our bodies. In the pages ahead we explain why we think we are on the cusp of a major transition that, revolving around oil, will dominate the investment picture for many years to come. And we tell investors exactly how they best can deal with this challenging new environment so as to come out winners. We hope you find it provocative, informative, and helpful.

Introduction to the Paperback Edition (2005)

We began writing *The Oil Factor*, the hardcover version, in 2002 because we had an urgent message for investors: oil prices are on the rise, the uptrend is long-term, and rising prices will come to dominate the economic and investment outlook. The book explains why oil supplies will become inadequate sooner than is generally understood, why demand will continue to rise, and how investors should respond.

Talk about catching a wave! Since then oil has been in the headlines constantly, prices have reached previously unseen levels, and cheap energy no longer is taken for granted. In other words, nothing has happened to change our views, and a lot has happened to confirm them.

In fact, rereading the book from today's vantage point, we would change barely a word, and that, actually, is pretty remarkable given how quickly things can happen in this world. Rather, we would simply cite further evidence that an oil squeeze is becoming ever more imminent and intractable. We also have discovered a handful of stocks that we would add to our portfolios. (Obviously, when it comes to buying, holding, and selling particular stocks, a book isn't the best medium for keeping up with the latest reports and seizing upon the newest opportunities. For investors interested in following our recommendations on an ongoing basis, we suggest subscribing to our investment letter *The Complete Investor* at www.complete investor.com; 866-833-2070.)

Here, in this introduction written for the paperback edition, we make just a few additional observations about oil and our future. We also suggest a variation on how to use oil prices as a stock market indicator. And we draw your attention to those additional stocks that will benefit in a big way from rising oil prices and growing oil shortages.

Capacity Limits

The first observation has to do with the world's oil capacity. Back in early 2002, $30-a-barrel oil was widely viewed as an anomaly. Most anyone with an opinion on energy—from Wall Street analysts to oil industry executives—believed oil prices were in a long-term trading range centered around $20 a barrel. This opinion rested on the assumption, which these optimists took as an article of faith, that worldwide oil capacity was sufficient to supply the world with all the oil it needed. True, much of that capacity was in the hands of the Saudis— but Saudi Arabia had always proved a reliable supplier of last resort. Moreover, even if you didn't have unlimited faith in the Saudis, there was always the emerging

Russian oil miracle: from the late 1990s on, Russian oil production and exports had been rising sharply.

But by 2004, it was becoming clear that a major change had occurred. Production increases in Saudi Arabia and Russia had become insufficient to offset burgeoning demand for oil from the rest of the world—from China, India, and the U.S. In mid-2004 the respected International Energy Agency issued a report stating that excess capacity in the world's oil markets was under one million barrels a day. That is a razor-thin margin: we live in a world in which a well-placed bomb could destroy a critical pipeline and lead to a loss of far more than a million barrels a day, a world in which a political dispute in Russia or Nigeria or Venezuela could easily cut off several times that amount. Clearly, by 2004, excess capacity had become woefully inadequate.

This brings us back to rising oil prices. The steep rise in oil prices (which, ironically, began in the wake of the U.S.'s toppling of Saddam Hussein) has been based on the emerging realization that the world simply does not have enough oil capacity to satisfy growing demand. Demand from Asia, as well as from the U.S., is not going to abate. The oil markets have simply recognized this reality, and they will continue to do so.

This means that oil prices are headed higher even if one of the points we develop in the book turns out to be overly pessimistic. In chapter three, "The Geological Lowdown," we make the case that energy supplies are within just a few years of peaking. Thus, we argue, even if demand were to slow, oil prices will keep rising.

It has become ever clearer, though, that demand *isn't* going to slow. So even if it turns out that there's more oil in the world than we think, enough to allow for some increases in supply for the next five, ten, maybe even fifteen years, oil prices still are headed higher. The lessons of the early years of this century are that even under the most realistically optimistic assessments of reserves, the demand for oil will outstrip the world's ability to increase supply.

But, you might ask, are we really giving the possibility of major increases in supply their due? After all, what about Saudi Arabia and Russia? Won't they be able to increase supplies dramatically? Isn't it possible that the early part of this century will prove to be just an anomaly, a short-lived bottleneck of an era during which Saudi and Russian oil was gearing up to come on stream?

Unfortunately, the answer is no. The reason is that, to the extent that Saudi Arabia and Russia can produce extra oil, it's almost inevitable that they will need to consume much of it themselves. In effect the world will be running in place.

Here's why. Let's give the oil optimists the benefit of the doubt and assume that both Russia and Saudi Arabia actually have tremendous reserves to develop. (This is by no means certain; in chapter three we explain why we think it very likely that all OPEC nations, including Saudi Arabia, have dramatically overstated their reserves.) To develop those putative reserves, however, will require a tremendous investment, billions upon billions of dollars. All this money pouring into their economies will go hand in hand with surging economic growth. And guess what—growth requires energy.

Let's look at Russia first. In the 1990s and early part of this decade, Russian GDP dropped by more than 30 percent. A good thing for the rest of the world, because it meant that Russian demand for energy fell by 20 percent. If, instead, Russian energy consumption had grown by just 2.5 percent a year during the same period, Russian consumption by 2004 would have nearly matched production, leaving almost no oil to import. But then, as energy production within Russia began to climb, so did energy consumption. We see no reason to think that Russian consumption won't continue to keep pace with production, because, as we said, rising production implies a rapidly growing economy.

Don't count on Saudi Arabia, either. Its internal consumption of the oil it produces has been in a sharp uptrend, the result of a host of energy-intensive activities that are part of the structure of the Saudi economy, such as desalinization, defense, and energy production itself—it takes oil to produce oil. Moreover, the uptrend occurred despite relatively tepid economic growth in Saudi Arabia. If the country embarked on a massive oil development program, economic growth would accelerate. In this energy-intensive economy, whenever growth has exceeded 4 percent, energy use has grown by 12 percent a year.

One way to look at this is to realize that if Saudi Arabia were to double its oil production capacity from 2004 levels—to 20 million barrels a day from around 10 million barrels a day, an extremely tall order—the country would need approximately half the additional barrels for its own internal use. That would leave just 5 million additional barrels a day for the rest of the world. And this wouldn't get the world too far; it represents less than three years' worth of growth, assuming the world continues to grow at around the same pace as in 2004.

The bottom line is that while oil may be volatile shorter term, longer term there are simply no scenarios—apart from a massive worldwide recession—that do not leave oil and energy prices in a sustained and powerful uptrend. It's basic economics and simple logic: as long as growth in oil supplies isn't keeping pace with growing demand, oil prices will continue rising.

To change this around, whether by curbing demand or boosting supplies, will require either highly unpleasant scenarios—cures that are worse than the disease—or time. A severe worldwide economic downturn would do the trick. So might conservation, eventually, but not overnight, and we probably don't have the luxury of time. As far as increasing supply, we indicated above why relying on developing more oil is surely a pipe dream. The ultimate answer—the development of alternative energies that can take oil's place—like conservation isn't going to happen quickly enough to head off higher oil prices. The upshot: oil prices are headed higher.

Our Oil Indicator

In chapter one, we explain why the best we can hope for is that oil prices rise in a gradual rather than sharp manner. The economy can deal with a sustained gradual uptrend, but a dramatic spike in prices would send it into a tailspin.

This brings us to our oil indicator, introduced in chapter two. It is based on the differing ways in which the economy, and therefore the market, reacts to oil price rises that are gradual vs. price rises that are abrupt. When the rise in oil prices is gradual, you should stress stocks that benefit from inflation. If prices rise too quickly, how-ever, you should focus on deflation beneficiaries. The level we chose as a demarcation line was an 80 percent year-over-year gain in oil prices. That is, when oil prices rise by 80 percent from year-earlier levels, the indicator suggests that investors shift from a portfolio weighted with inflation plays to one geared to deflationary times.

Some readers, however, felt that we had attempted to make our oil indicator too precise—too all-or-nothing. And we would be the first to admit there is nothing magic about the 80 percent level. In fact, we noted in the book that other levels also would work, but that based on the historical data, the 80 percent cutoff makes a lot of sense. The overriding point, though, is that there is no better bear market forecaster than rising oil, and the larger and quicker the gains in oil, the more likely a recession and the more likely a bear market.

There are, however, alternative ways to play the oil in-dicator other than switching wholesale between a defla-tionary and inflationary portfolio. For any readers uncomfortable with such an all-or-nothing rule, we offer here some other approaches. (You can refer back to these when you get to chapter two.)

One valid approach would be to have two portfolios and weight them according to how oil is doing. Given the strong correlation between oil and the market, virtually any rule you pick—and we tried to make this clear in the book—will work. Suppose, for example, oil prices rise 50 percent over 12 months (as was the case in August 2004). At that point you might weight the two port-folios equally; going forward, you could subtract 10 per-centage points from the inflationary one and add 10 percentage points to the deflationary one for every 10 percentage-point gain in oil. And conversely you would subtract from the deflationary portfolio when oil dips by 10 percentage points. We suspect that this approach is unlikely to work as well as following our strict 80 percent rule—but it would allow for more gradual and less jarring changes. So when you get to chapter two, keep it in mind.

A Few More Stocks

Another approach is to invest more heavily in stocks that are suitable for both inflationary and deflationary portfolios. One group that fits the bill is alternative energy companies. In chapters six and seven, we discuss a wide range of alternative energies—some of them already in use or well developed, others more remote. The ones likeliest to have the most immediate impact on helping us meet our energy needs include nuclear energy, wind energy, and tar sands. Companies that help promote energy efficiency also belong in the mix. And in the past two years we also have come to give more credence to the role that liquefied natural gas is likely to play.

Unfortunately, when we first wrote the book, we could find only a few companies that were significant players in these areas, and for the most part they were very sensitive to economic conditions. Despite their stake in critical energies, they had other businesses that were very sensitive to economic conditions and hence were not suitable for a deflationary portfolio.

More recently, though, we have found a few more companies in the above areas and have had second thoughts on others that are both big players in alternative energies and also wonderful hedges should the economy falter. In other words, these are all-weather stocks that can go in any portfolio and thus help bridge the gap between our inflation-oriented and deflation-oriented portfolios.

Here we present four companies not included in our original portfolios but that we now strongly recommend as additional alternative energy recommendations in both our inflationary and deflationary portfolios. They are FPL Group, Entergy, Exelon, and Canadian Oil Sands, and we'd weight them equally in both portfolios. We also discuss one additional stock that belongs just in our inflationary portfolio: Air Products and Chemical. Below is a brief discussion of these stocks and our reasons for liking them.

In chapter six, we explain why fission-based nuclear energy won't be a long-term solution to our energy needs: the laws of physics dictate that disposal of nuclear waste on a large scale is simply not feasible. Still, nuclear energy will temporarily help fill the gap. Exelon is the nation's largest nuclear utility. Moreover, the company's entire nuclear fleet is merchant or unregulated. We expect nuclear energy's cost advantage will continue to rise in the years ahead. Of course the kicker for any major nuclear energy provider is the potential to build additional facilities or to add to capacity at existing facilities, and we think this is likely. For Exelon this will mean higher margins and higher usage, which along with the potential for expansion would result in double-digit profit growth. With Wall Street projecting much slower growth, double-digit growth would translate into a sharply higher P/E along with higher earnings. Because of the stock's high and secure current yield, it also will shelter you during periods of economic weakness.

Entergy is the nation's second largest nuclear utility and like Exelon is an exceptionally well run company. The major difference is that some of Entergy's nuclear-generating capacity is regulated, which means you won't get the full effect of margin expansion as fossil fuel prices rise. But you will get even more downside protection in the event of a recession and will also have plenty of up-side exposure to nuclear energy. As with Exelon, long-term growth is likely to surprise Wall Street on the upside.

Another energy source destined to gain traction in the years ahead is liquefied natural gas, or LNG. We short-changed it in the book, but since then we've done a lot of research, and we now estimate that over the next fifteen years LNG has the potential to add the equivalent of about five or six million barrels a day to the world's energy supply. That won't solve the world's energy problems, but at least it will make a dent. A company that plays a critical role in LNG is Air Products and Chemical. It makes heat exchangers that are vital components in converting natural gas into liquefied form that can then be shipped. Currently heat exchangers account for only a few percent of Air Products' revenues but in the years ahead this technology could add several percentage points to this high-quality company's growth rate.

Another reason for buying Air Products is that it also produces hydrogen. While the amount is not enough to transform the U.S. into a hydrogen-based economy, hydrogen is a vital component in refining alternative fossil fuels, including the vast Canadian tar sands. With nearly

180 million barrels of oil sands reserves, these tar sands are the second largest source of fossil fuels in the world. The catch, which we discuss in chapter thirteen, is that it is both difficult and expensive to mine these sands. Most frustrating, mining them is highly energy-intensive, meaning that as oil prices rise, so does the cost of extracting energy from tar sands.

Still, we expect tar sands to occupy an important alternative energy niche. One stock that, thanks to its high yield, is appropriate for our inflationary and deflationary portfolios alike is Canadian Oil Sands. While we don't want to get too technical, it is important to know that the company is a trust, meaning it has to pay out a large portion of its earnings as income. For that reason the stock's yield will remain much higher than that of most other energy producers, affording a fair amount of downside protection and income in the event of a recession. At the same time, the company has a sure stake in tar sands as they become a more significant source of energy. Double-digit growth in both earnings and income is likely over the next decade.

Finally there is wind, which is the most promising alternative energy. The two most significant wind generators in the U.S. are General Electric, which we recommended in chapter thirteen, and the Florida-based utility FPL Group. FPL is one of the best-situated all-weather stocks around and belongs in just about every portfolio all of the time. The company has two businesses. The largest, approximately 85 percent of revenues, is a regulated utility operating in one of the best demographic areas in the country, Florida. This regulated business is a staid but steady generator of income. If this were the sum total of FPL, it would have no place in an inflation-oriented growth portfolio. The kicker here is that FPL, through its smaller deregulated businesses, has become one of the best-situated alternative energy companies in the country.

What makes this an exceptional risk/reward opportunity is that no one on Wall Street has noticed the tremendous growth potential of the deregulated businesses. Thus Wall Street continues to estimate overall growth for FPL based on the modest levels by which its regulated businesses are growing. Yet revenues and income from the deregulated businesses have been growing considerably faster, making them an ever bigger part of the company as a whole. And the best may lie ahead, because FPL is by far the nation's largest wind generator. (General Electric, which manufactures wind turbines as well as produces wind power, is the largest integrated wind company.)

Several highly reputable scientific studies, presented in chapter seven, argue convincingly that wind has become the most cost-effective way of generating electricity. Moreover, most of these studies were conducted when prices for natural gas and other fossil fuels were far below current levels. Indeed even in Britain, where in 2004 natural gas prices were roughly 50 percent lower than in the U.S., wind electricity is nearly comparable in cost to gas generation.

We expect wind to continue to generate outsized growth—enough to propel FPL's overall growth well into double digits. There is an additional kicker as well. FPL's unregulated business also has an emerging and potentially very large stake in LNG, with agreements securing its equity participation in a major LNG terminal and

LNG-related pipeline that will serve the Florida area. Implementation of these projects would secure FPL's position as a major LNG player, a position that would further add to long-term growth.

Now on to the rest of the book, where we lay out in detail our case for a continuing uptrend in oil prices and tell you what to buy and what to avoid in order to protect yourselves and profit. Even in the turbulent times that lie ahead, the upside potential for investors is enormous. But the risks are becoming greater, too. To dodge the bullets and latch on to the winners, you need a clear-eyed understanding of what lies ahead. We've done our best to offer it to you in the pages that lie ahead.

Introduction

A t various moments in the past thirty-plus years, oil has thrust itself front and center into our consciousness. The first oil shock came during the Arab oil embargo of 1973, when oil prices rose precipitously and filling up our tanks at the local gas station became a major ordeal. A few years later during the revolution in Iran, and then in 1991 prior to the first Gulf War, prices also soared. During these times, everyone fretted about oil.

When oil prices came down, though, as they always seemed to do, most of us once again were happy to take oil for granted. Yes, we knew it was a finite resource that eventually would run down—but surely that day was a long way off. And yes, we had become dependent on an unstable and often hostile part of the world, the Middle East, to supply us with the lifeblood of our economy—but surely it was in those countries' interests to continue to do so.

During those years, numerous Cassandras (and remember, while the term "Cassandra" is often used in a derisively negative way, the original prophetess of that name was absolutely on target in her gloomy rantings), including President Jimmy Carter, warned that the U.S. needed to begin planning sooner rather than later for the day when oil would become less available/wildly expensive. Many were motivated by environmental and geopolitical concerns as well as by the realization that at some point the oil we needed simply wouldn't be there.

From the perspective of the early twenty-first century, it's clear that those pushing for the meaningful development of alternative energies were almost eerily prescient. Reading their speeches and white papers and op-ed pieces, it is hard not to feel incredibly frustrated that no one paid more attention. If the political will had existed during the past three decades to do more than pay lip service to alternative energies—if by today, for instance, as Carter had proposed, wind energy accounted for a significant portion of our energy usage, or if we had launched large-scale programs to develop more efficient solar cells—well, it's hard to overstate how much better off we'd be, and the world as well.

♦ ♦ ♦

We didn't, however, and nothing can turn back the clock. We are where we are and have to deal with it as best we can. In the pages that follow we offer an investment road map to where that is and to what lies ahead. But if we started this introduction by talking about oil, it was for good reason. Or to be more precise, it was for two good reasons.

First, we believe that we are on the verge of a historic transition away from relying on oil as our primary fuel.

This transition won't be something we choose. Rather, it will be forced upon us by the fact that oil supplies are peaking—within a very short time, oil producers simply aren't going to be able to produce as much oil as the world needs. This will be the key big-picture trend of the 2000s, and it will lead to a long-term uptrend in oil prices. Oil shortages and rising oil prices will be the order of the day, and they will play havoc with the economy. The upshot will be turbulence and uncertainty.

Most of the time, inflation will rule, as rising energy prices spur overall price increases. And this represents a sea change in our economic environment. Inflation has been remarkably quiescent in recent times. During the 1990s and the first several years of the twenty-first century, inflation averaged under 3 percent. But the days of stable prices are just about over. In tandem with the onset of a new era of shrinking oil production, we are also entering a new era in which inflationary pressures will mount.

It may seem quixotic that we are focusing on inflation when lackluster economic growth and the threat of deflation dominated discussion about the economy as recently as 2003. But these concerns are likely to be overwhelmed by rising oil prices. One thing to keep in mind is that as oil prices move higher, energy costs will become an increasingly bigger part of the economy, and the same percentage increase in oil prices that once had a negligible economic effect will suddenly be significant. It's as if you had a metal cup that weighed a pound and that had a single penny at the bottom. Even doubling or tripling the number of pennies wouldn't affect the overall weight much. But once you get to a certain point, the same percentage increase in the number of pennies would have a very noticeable effect—the cup suddenly would feel a lot heavier. A similar sort of thing will happen as oil prices rise. Today oil is still a relatively small part of the economy, but as it grows more expensive, the effects on the economy of rising oil prices will become more and more pronounced.

◆ ◆ ◆

All in all, oil prices are likely to rise to triple-digit territory—$100 a barrel at a minimum, and probably higher—by the end of the decade and possibly sooner. Inflation ultimately is likely to reach levels well into the double digits.

But we aren't dismissing deflationary concerns entirely. It is entirely consistent with rising energy prices and rising inflation that at times the basic inflationary uptrend will be interrupted by deflation or, what will have the same effect, by fears of deflation. These contrary but symbiotic forces—inflation and deflation—will seesaw back and forth in chaotic fashion, with inflation generally ascendant but not always. The interactions of these two forces will establish the parameters of the entire investment environment for the next decade or longer.

It's worth making one final point here, one that is a bit complex but that we explain in more detail later on. Even if inflation rises less quickly than we expect, we still almost certainly will be facing what we term an "inflation-like" environment, which will have almost identical repercussions for investors. By this we mean a situation where inflation, though not particularly high in absolute terms, is above the level of interest rates. It's not quite the same thing as inflation, but it results in the same type of relative performance by so-called inflation hedges.

◆　◆　◆

For investors who don't have a clear understanding of these trends, the coming years could be a true nightmare. Investments that seem to be on track when inflation is in the driver's seat may suddenly tank when deflationary fears erupt, and vice versa. Moreover, in contrast to the 1990s, you won't be able to solve the problem by simply buying a representative index of seemingly safe, conservative stocks, for these are exactly the stocks that will lose ground to inflation.

In addition, in the years ahead, a buy-and-hold strategy won't do. If you buy the right stocks, those that tap into the prevailing economic winds, you'll make good gains. But to hold on to those gains, you will need to be more proactive—to be willing to shift into inflation beneficiaries as inflation takes over and into deflation hedges when deflation seems the main threat.

And this leads to the second reason we opened by mentioning oil. For oil not only is the prime cause of this turbulent transition we're entering—it also, as it happens, is your best guide to getting through these difficult years.

As we'll detail in chapter 2, "Our Amazing Oil Indicator," over the past three decades, during good times and bad, bull and bear markets alike, oil prices have been the single most reliable guide to stock market performance. When oil prices have risen too abruptly, the stock market as a whole has been a losing proposition. When prices have dropped, or risen only modestly and gradually, stocks have soared.

In the coming years, it will be even more crucial to pay attention to oil. Even as oil supplies falter, oil prices will remain the key to stock market performance. Keeping an eagle eye on oil prices will be the one thing that can keep investors on the right side of market lurches. We explain exactly how to use our oil indicator to know when to emphasize inflation hedges and when to focus on deflation hedges.

◆   ◆   ◆

What are these inflation and deflation hedges? While part I lays out the big picture, in part II we tell investors exactly which groups and stocks will be winners in each type of environment. Mainly we delineate a carefully selected group of investments that will benefit from energy shortages and inflation. As inflation heats up, these investments will shine—even as most other groups are losing ground.

Above all, investors—in stark contrast to what was a winning strategy in the 1990s—will need to focus on all classes of real assets. These include energy stocks, precious metal stocks, and real estate. We explain why these categories of investments, which are directly leveraged to inflation and thus are the quintessential inflation hedges, will be must-haves in the coming years and how to pick the best of them. But other groups also should be core holdings in nearly every portfolio, including defense stocks and alternative energy stocks.

Even when inflation is in charge, though, all investors still should own some deflation hedges, such as zero coupon bonds. That's because, as we said, the coming period will be unusually turbulent. Our oil indicator is the best indicator we've ever seen, but that doesn't mean it's perfect—no indicator is. Some deflation insurance is necessary at all times for all investors. Anytime our indicator tells you that deflation is gaining ground, investors should strongly increase the proportion of deflation-leveraged investments they own, deemphasizing inflation hedges until the indicator turns around again.

◆　◆　◆

Bob Dylan said it best: "The times, they are a-changing." In the 1990s, it was easy to be an investor. You could throw darts at stock listings in the *Wall Street Journal,* you could buy the big safe indexes and make a fortune, never losing a night's sleep. Then came the relentless fall in stocks and the rocky times of the early 2000s. These more recent years are the prelude to the years ahead. If you're still hoping for a return of the 1990s, you're going to be disappointed. Oil shortages and rising oil prices will make sure of it. It isn't going to be easy.

But this doesn't mean it is hopeless. If you understand the trends and zero in on the right investments, using our oil indicator as your guide, you should outperform the overall market in a big way. In fact, you'll likely do better than the great majority of so-called professionals. This will be a period that will exaggerate the differences in performance. Our book is designed to show you how to be on the side of the winners.

Part I

The Big Picture

If you're like many investors, you want to cut to the chase. You have limited patience with theories and explanations—you just want stocks that will make you lots of money. You want names, symbols, prices, and targets that are temptingly high but firmly realistic.

Well, sorry. For the next nine chapters, we offer a ton of explanation. You can skip them if you want and head right to part II, where we give you those names and targets, but we think it would be a mistake. We are ardent advocates of the view that all investors should be as well-informed and astute as possible about the world that lies ahead. We want to make you so aware of the underlying key trends in the world, so alert to what will make stocks move up or down, so knowledgeable about how to evaluate companies in the particularly challenging circumstances that lie ahead, that you will feel utterly in control of your own financial destiny. Whether you manage your own investments or hire someone else, you will know what to look for in an investment, what questions to ask, what to avoid. The following pages are meant to give you that grounding.

In them we do our best to build an airtight case for what the economy will be like in the next decade and possibly well beyond. Our case has three main elements that are closely intertwined. First, we demonstrate the enormous importance of oil to the economy and the stock market—why oil counts. Second, we show why oil (and natural gas) prices have broken out of long-standing trading ranges and entered into long-term uptrends. And third, we show why inflation pressures will erupt in conjunction with a hovering deflationary threat. Once you understand these three elements and how they will work together to reshape our entire economy, you will have at your fingertips most of what you need to maximize your investment results.

If oil weren't so critical to our economy, rising oil prices wouldn't be a big deal. No one would care, for instance, if chocolate prices entered a sustained and sharp uptrend. Well, actually a lot of people might care, but still it wouldn't affect the economy and stock market as a whole. But an uptrend in oil does. To understand why this is so, we sketch the history of oil in relationship to our economy and financial markets. And when you break down this history carefully, as we do, you can see that because of oil's overriding importance to the economy, you can use changes in oil prices in a very specific way to tell you when to get out of the stock market and when to get back in. In fact, it's almost the only guide you need.

In showing why oil prices will rise to new heights we look at both the supply and the demand side, tackling such arcane but important topics as the geology of oil and the consequences of debt. We also discuss alternative energies and conservation, to see how they might affect oil prices and to gain a clearer understanding of the long-range trajectory of the energy crisis and its eventual resolution.

It all begins with and revolves around oil. It's that simple and that important. Once you understand all the implications of this reality, you will be one hundred percent ready to move on to part II, in which we get down to the nitty-gritty of what to buy and what to avoid in the new era of oil-driven inflation that lies ahead.

CHAPTER 1

Thirty Years of Oil

T he most important events in history, the ones that will have the greatest impact on our lives for years to come, often slip by unnoticed at the time. Go to a library and scan issues of the *New York Times* from the fall of 1960. What was making news in that presidential election year, apart from coverage of the Kennedys' glamour and Nixon's five-o'clock shadow? Two tiny islands called Quemoy and Matsu; and Nikita Khrushchev; and our fledgling space program.

Definitely *not* grabbing headlines, in an era when oil was priced at under $2 a barrel and the U.S. satisfied around 70 percent of its oil needs through domestic production, was the decision, in September 1960, by five countries—Iran, Iraq, Saudi Arabia, Kuwait, and Venezuela—to form a loose coalition called the Organization of Petroleum Exporting States, or OPEC. But the ultimate repercussions of that event have been massive. In fact, as we will detail below, it is no exaggeration to say that OPEC, which gradually expanded to include Qatar, Indonesia, Libya, the United Arab Emirates, Algeria, Nigeria, and Ecuador, has become the single most important determinant of the health, or lack thereof, of both our economy and our financial markets.

Ten years later, another oil-related economic milestone also got little attention. In 1970, U.S. domestic oil production, which up until then had been consistently rising, embarked on a decline, one that has continued ever since. To the extent that anyone noticed it at all, it was viewed as either a temporary anomaly or as simply no big deal. Like the formation of OPEC, however, the decline in domestic oil production has been of critical importance to the economy and to investors.

Oil is key to all we do, to every facet of our economy. Or to put it more precisely, energy is key, and for now, because of a long-term failure spanning administrations of both parties to develop alternative energies, energy means oil. Our need for oil, our growing appetite for this critical resource, is the prism through which it is essential to view all that is happening in the world today and all that will occur tomorrow. This is true for all of us, citizens in general. And it is true in an even more specific way for investors who want to understand what is likely to happen in the financial markets in coming years and what they need to do to protect themselves and profit.

Above all, it's essential for investors to grasp, intellectually and viscerally, the following realities:

• Since 1973, the price of oil has been the single most important determinant of the economy and the stock market. Sharp rises in oil prices have been deadly for

◆

the economy and the stock market, while steady or declining prices, or even prices that increase only gradually, have led to good times. For investors, it's what we dub your "desert island, one phone call" indicator. If you can know only one thing about the world, make it the direction of oil prices over the preceding year, and you'll do better in the stock market than almost anyone else following any other indicator, from interest rates to corporate profits. This has been true for the last three decades, and it will remain true throughout the early part of this century—until we kick our oil habit and develop and switch to viable energy alternatives. In chapter 2 we give you specific guidelines for using this exceptional indicator.

- Oil prices are a determinant over which, for the past thirty-plus years, we have more or less ceded control. In other words, through good times and bad, we have exercised little real control over our own economic fate.

- Finally, the situation is about to shift from bad but acceptable to worse, because, as we'll detail in chapter 3, for all practical purposes the world is running out of economically extractable oil. This puts us more than ever at the mercy of the very few nations with significant untapped reserves—Saudi Arabia and to a lesser extent Iraq, Iran, and Kuwait. Over the long term it's clear that the only viable solution is to free ourselves from our dependence on oil entirely, by shifting to other forms of energy. But in the meantime, we are trapped in a tricky and dangerous present, in which we need to ensure that we have the oil we need to keep the economy going while we seek to develop alternatives on a meaningful scale. This doesn't mean that the economy is doomed or that there aren't significant profits to be made in the stock market during the tumultuous transition that lies ahead. It does mean, though, that you have to know what to look for—in particular, that you need to watch oil, tracking the direction of oil prices at any given time and then tailoring your investments to fit what oil dictates.

Below and in following chapters we'll discuss these three key points in detail. We'll explain where to go to find oil prices at any given time, how oil ties in with other salient economic realities, such as super-high levels of consumer debt, and what this means for the short-term and long-term outlook for the economy and stocks. In particular, we explain why the upshot is likely to be high and rising inflation accompanied by the ever-present risk of deflation, a volatile combination that will transform the investment environment. And we'll tell you exactly how to use trends in oil prices to catch each investment wave and to profit whether stocks are going up or down.

In this chapter, though, we'll look at the recent past— at the decisive though often surprisingly overlooked hold oil has exerted over our lives for the past thirty years.

A Brief History of Oil Prices

It is striking, if you look back to the years since 1973, how closely changes in oil prices have mirrored both the economy and the stock market. During this period, rising oil prices have always preceded economic downturns and falling stocks. Falling oil prices have always led to economic upturns and rising stocks. It's that simple, that predictable. And there is good reason for this strong correlation. For a long time, oil has been our major energy source, and economic growth depends on the availability of energy as much as the growth of a child depends on the availability of food. When energy is available at low prices, the outlook for growth is good, and stocks go up. When energy prices go up, growth becomes harder to achieve, and stocks go down.

Now, you might point out that our economy was running on oil long before the 1970s, and that's true. But in those earlier years, no one ever thought much about oil prices. Oil was just there, like air and water—other commodities we gave a lot less thought to back then. It was cheap, it was plentiful, and it was dependable. Businesses could count on getting all they needed, and so could consumers.

In the fall of 1973, that age of innocence vanished forever, as a result of the 35th OPEC conference, which began in Vienna in September and ended in October. This event transformed our economic landscape and forever changed how we think about oil. During that conference OPEC imposed restrictions on oil exports. In so doing, it engineered a 70 percent increase in oil prices, which rose to the then unheard-of level of more than $5 a barrel. In December the cartel met again, this time in Tehran, and took even more drastic action. Protesting U.S. support for Israel in the 1973 Yom Kippur War, it temporarily embargoed oil exports altogether. By early 1974 oil prices had jumped to more than $7 a barrel, more than 130 percent above levels that had prevailed just a few months earlier, in mid-1973, and, indeed, for the entire preceding decade.

OPEC had done what the Soviet Union, throughout the Cold War, had failed to do—demonstrated not by threats but by action our vulnerability to forces over which we had no control. The cartel continued to flex its muscles, and oil prices continued to rise throughout the 1970s, in a steady but constrained uptrend. Then the situation abruptly worsened. Propelled by the overthrow of the shah of Iran and the Iran-Iraq War, oil prices soared. By the end of 1979, oil—which had averaged a shade over $10 a barrel in 1978—was more than $18 a barrel. And by early 1981, prices had reached nearly $40 a barrel.

Then the pendulum shifted again. The reason: the West had reacted to the rise in prices by cutting back on its oil use through a combination of conservation and the development of other energy sources. Heavy investments in nuclear power and the development of coal and of oil fields outside of OPEC's reach began to pay dividends. The combination of lower demand and increased supply drastically reduced OPEC's ability to control prices. As a result, oil prices were in nearly free fall during much of the 1980s. From their highs of nearly $40 a barrel early in the decade, they plunged to a low of near $10 a barrel in 1986.

By mid-1987, though, oil prices rose significantly once more, though getting nowhere near their previous highs. The reason for the rise was, once again, OPEC, which, alarmed by prices in the single digits in 1986, had reined in production slightly. For the next three years oil prices fluctuated between the mid- and high teens without any obvious trend.

You probably remember what happened next. On August 2, 1990, Iraqi president Saddam Hussein ordered his army to take over Iraq's nearly defenseless neighbor, Kuwait, a Muslim country whose population numbered only about two million. Though Kuwait was barely a speck on the map and shared virtually no Western values, it took just twenty-four hours or so for the UN Security Council to order Iraq to withdraw. Why such a fuss over Kuwait? Oil, and a lot of it. Kuwait, a founding member of OPEC, was one of the world's largest oil producers. It was clear that Saddam had only one thing in mind, and that was possession of Kuwaiti oil. The threat of so much oil in the hands of an unpredictable dictator was enough to make the West act, and act decisively.

As the Gulf crisis unfolded, the oil markets responded by driving up the price of oil by over 50 percent in just a few weeks. A number of catastrophic scenarios were being bandied about. The most alarming one was damage to Saudi Arabia's oil fields. Many analysts argued that a desperate Saddam would send missiles to every corner of the Middle East, inflicting untold economic damage in the West. Saudi Arabia was considered the most likely target because it was the world's largest producer of oil. Oil prices soared to above $30 a barrel.

Once the shooting started, however, in mid-January 1991, it became clear that Saddam offered no real resistance and lacked the means to damage the Saudi fields. Though the Kuwaiti fields were left aflame, the Saudis had more than enough capacity to make up for the short-fall until the Kuwaiti fields were repaired. As a result, by early 1991 oil prices had fallen back to the high teens. For most of the rest of the decade, prices remained well under control. Oil hovered near $20, with the average price a shade above $19. Moreover, as figure 1a, "Oil in the 1990s," shows, whenever oil peeked above $20, it quickly backed down. Remember, too, that $20 oil in the 1990s was comparable, after adjusting for inflation, to $15 oil in the 1980s.

It wasn't until the end of the decade that oil began to display any volatility at all. In 1998 OPEC, because of an internal battle over market share, turned on the taps full blast. The extra oil hit the markets just as the Asian economies were entering a serious swoon. At their lows in 1998 oil prices dropped to $10 a barrel. OPEC became more disciplined in 1999, and oil prices recovered all the way back to the mid-20s on the heels of surging economic growth. Oil finished the decade at about $25 a barrel, not alarming but still the highest level since the Persian Gulf War of 1990–91.

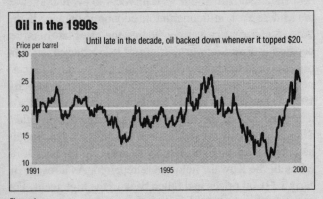

Oil in the 1990s

Price per barrel   Until late in the decade, oil backed down whenever it topped $20.

Figure 1a

Oil, the Economy, and the Stock Market

During the years between 1973 and the turn of the century, while oil prices bounced up and down between a low of around $10 a barrel and a high of near $40, we experienced five recessions and several periods of strong economic growth. Meanwhile, as far as stocks went, we've had bear markets, in which stocks dropped nearly 50 percent, and some prolonged and glorious bull markets. Who says life is dull?

The point is that if you looked at when the economy went into a tailspin and when stocks tanked, you'd see that these periods were always preceded by rising oil prices. By the same token, falling oil prices preceded economic good times and strong financial markets. It's like those acetate overlays in books depicting the human body, in which a sheet depicting branching blood vessels lies neatly over a sheet showing the skeletal frame. If we had such overlays, we could fit economic and market trends snugly within changes in oil prices.

Let's briefly look back at the economy and market during those same thirty years since 1973 and see just how their ups and downs intersected with trends in oil. It is no coincidence that the period 1973–82—which saw oil prices rise from below $5 a barrel to nearly $40 a barrel—was one of the most turbulent in U.S. economic history. Inflation soared into double digits, and the economy experienced three recessions. Moreover, during those years, economic dogma was rewritten. Until then there was a well-established relationship between economic growth and inflation. When growth was strong, inflation would pick up; when growth was reined in—by rises in interest rates and a tighter money supply—inflation would fall.

This relationship, which had long been a reliable road map for economic policy, was blasted out of the water by events in the 1970s. Because of our lack of control over oil, inflation and recession were no longer mutually exclusive. As OPEC engineered rises in oil prices, prices rose across the board, even in the face of a slowing economy. In the 1970s, a new term entered the economic lexicon: "stagflation," a combination of stagnant growth and high inflation. At times we didn't know which of these enemies to fight first. In 1974, thanks in part to Alan Greenspan, then economic adviser to President Ford, Americans were urged to wear "WIN" buttons, standing for "whip inflation now." Those buttons were quickly discarded when it turned out that a more devastating enemy was the recession that had started in 1973 with no one noticing.

Well, not exactly nobody—the stock market clearly noticed. The year 1974 was one of the worst ever for stocks. Between the start of the oil embargo in December 1973 and their low in 1974, big-cap stocks as measured by the S&P 500 plunged by over 30 percent, while smaller-cap stocks suffered even more damage. With the exception of gold stocks, nothing was spared.

For the remainder of the 1970s, with oil uptrended but not abruptly so, and inflation seemingly on a permanently higher plateau, stocks treaded water. But then the dramatic rises in oil prices at the end of the decade and early in the 1980s—the result of political turmoil in the Middle East—hit the economy and the stock market hard. Between 1980 and 1982, the economy suffered through two recessions as unemployment soared into double digits. Stocks bobbed and weaved and in the end generated negative real returns for the period.

Then, however—as oil prices subsided and then dropped sharply throughout the 1980s, from their high of near $40 a barrel to a low of around $10 a barrel in 1986—a striking turnaround occurred. In the summer of 1982 a great bull market began. The Dow rose from a low of 780 in August of that year to a high of 2700 in 1987, and bonds soared as well. It was one of the strongest and longest bull markets ever. The economy was in high gear as well: economic growth was steady and strong, while inflation was decisively tamed, falling from the mid-teens at the start of the period to just a trace above 1 percent for the broad-based consumer price index (CPI) by the end of 1986. In short, it was an economically ideal period that was the mirror image of the 1970s. And it was made possible by well-behaved oil prices.

This period of nearly unmatched prosperity basically brought about the end of the Cold War, by giving us the ability to build up our defense establishment to the point where the Soviets simply could no longer compete. The triumph of capitalism was, seemingly, a great victory for the U.S. and the West, and not surprisingly it dominated the headlines and op-ed columns of the time. But it's clear in retrospect that everyone failed to appreciate the underlying dynamics that made our triumph possible. Oil was the silent dignitary at the table, the one that secretly held all the cards. The correlation between oil and economic prosperity was powerful and it was there for all to see, but no one was paying attention. The issue that is coming to a head now—that will determine our fate and possibly even our survival in the twenty-first century—was nowhere on our radar screens at a time when we more easily could have done something about it.

In 1987, we briefly experienced the flip side of the oil/prosperity correlation. As oil prices began to rise again, into the mid-teens—low by the standards set in the 1970s but high compared to the benign levels of the 1980s—other commodities rose as well. Inflation climbed back to above 4 percent, and the dollar began to tumble.

Predictably, stocks, which had been soaring for five years, began to falter during the summer of 1987 as well. The wavering culminated in one of the worst market sell-offs in history. On Black Monday, the Dow Jones Industrial Average plunged more than 500 points, and for the month stocks fell by more than 30 percent.

Staggering as the crash was, it actually marked the only time since 1973 that a sharp rise in oil did not end up triggering economic recession. The reason: because inflation was at still-manageable levels, the Fed had the leeway to pump sufficient money into the economy to ward off a downturn. The drop in 1987 turned out to be just a temporary dip on the road to even greater prosperity and much higher stock prices. For the three years following the 1987 crash, with oil prices essentially a nonevent, the market was in high gear. By their highs in 1990, stocks not only had recovered the ground lost in the 1987 crash but were more than 20 percent above their 1987 highs. All was right with oil, all was right with the world, and investors were making money hand over fist.

This euphoric period lasted until Saddam's invasion of Kuwait sent oil prices spiraling up by over 50 percent in just a few weeks, to more than $30 a barrel. The economy began its first recession in almost a decade, and stocks were pummeled.

As noted above, though, it quickly became clear that Saddam lacked the ability to cripple oil production. And as oil prices collapsed back to the high teens, the recession ended. Stocks once again embarked upon a bull run—and this one was to be a bull run for the ages.

Between 1991 and 2000, with oil prices remaining well under control, stocks staged one of the greatest rallies any financial market has ever seen. If you had invested in the S&P 500, say, by buying the Vanguard 500 Index Fund, in January 1991, you would have gained on average 20 percent a year for the next nine years. To put it differently, a $10,000 investment would have turned into more than $50,000. And because those nine years were ones of low inflation, your gains were mostly real gains in terms of their actual purchasing power. Moreover, as everyone knows, while the market as a whole was thriving, the tech sector, especially as the decade drew to a close, was on a real tear. All in all, these were unforgettable years, in which growth seemed to be on an ever-rising trajectory while inflation remained a virtual no-show.

Not the Same Old Oil Story

You may have noticed that in detailing the history of oil prices, the economy, and the stock market, we began our narrative in 1973 and seem to have halted it somewhere around 1999. You might wonder if on some psychological level this represents wishful thinking, a desire to linger in that relatively trouble-free period forever. Maybe so, but there is more to it than that. For toward the end of 1999 we began to move on to a new paradigm involving oil, a qualitatively different situation that requires that we look at the years since then separately.

You'll recall that in the 1990s, oil prices were stable, until 1998 averaging just a touch below $20 a barrel and quickly coming down anytime they ventured above the $20 level. Then they fell, as OPEC miscalculated and raised output just as the Asian economies were tanking. Once OPEC got its act together, prices rose once more, supported by the surging economic growth of 1999.

And then something out of the ordinary happened. Oil kept rising. By the third quarter of 2000, even though worldwide economic growth had begun to slow and the stock market was retreating, oil climbed above $35 a barrel, more than three times the lows of late 1998.

With the advent of winter and the threat of cripplingly high home heating oil prices, oil entered our consciousness, becoming a front-and-center crisis for a while. President Clinton authorized the release of oil from the Strategic Petroleum Reserve (SPR), the first time this emergency reserve had been used in peacetime. The only time prior to 2000 that oil had been released from the reserve was during the 1990–91 Gulf War. Clearly, the decision to use the SPR to curb soaring energy prices in the winter of 2000, whatever its political motivations, was an acknowledgment of the fact that oil prices were so high as to constitute at least a mini-crisis.

Why *didn't* oil prices come down in the early 2000s, when, as the result of slower worldwide economic growth, demand for oil was, if not lessening, growing at a very sluggish pace? Economic growth during those years was just 2.5 percent, about 20 percent slower than the average rate of growth during the preceding fifty years.

In fact, other than the recessions of the early 1980s and the 1990–91 period, it was the slowest-growing four-year period in postwar history. During those earlier recessions, oil prices dropped sharply. Indeed, until 1999, oil prices always fell during periods of economic weakness, rising only when demand was rising or as a result of ephemeral political disturbances. In other words, for most of the postwar period, oil prices were demand-driven.

But in 1999–2002, even though demand was weak, prices remained at historically high levels. This was due to a momentous transformation that had occurred in the dynamics of the oil industry: oil prices had become supply-driven. And the reason that had happened was that for the first time since its formation, OPEC was no longer just another player, albeit an important one, in the oil arena. It had become the controlling player.

Between 1982 and 1998, the oil-producing world outside of OPEC had been able to increase oil production enough to accommodate economic growth. In 1999, however, Britain and other non-OPEC oil producers hit the point where any increases in production were minimal. The only area outside of OPEC capable of meaningful increases in production was the former Soviet Union (FSU). And increases by the FSU were little more than sufficient to satisfy the world's marginal increase in demand.

To put it differently, by 1999 the world was using all the oil that producers outside of OPEC not only were generating but that they were capable of generating. Supply was barely keeping up with demand. The upshot: as the world steps up its need for oil—and as we'll explain, it is a given that it will, unless we enter a protracted worldwide recession—the only countries that can supply the extra barrels are OPEC nations.

The result is that since 1999, OPEC, through relatively small changes in its production quotas, has been able to exercise almost complete control over oil prices. Oil demand in the world has overtaken the ability of all but a small group of nations to supply the oil that is essential for world economies to survive.

One key question is whether this is a temporary imbalance or a new and more permanent reality. As we'll discuss in chapter 3, all the evidence points to it being the latter. Nothing the West can do—other than end its reliance on oil altogether by developing alternatives on a large scale—will free it of its dependence on Middle Eastern oil. And nothing other than the development of alternative energies (or permanent recession/depression, an unacceptable outcome) will forestall oil soaring to ever higher levels, though possibly with temporary periods of retrenchment. The tragedy is that because we have waited so long, nothing but soaring oil prices will push us to develop those oil alternatives on a grand enough scale.

So there you have it—an overview of the fateful dance between oil, the economy, and stocks. Next chapter we look in more detail at the relationship between oil and the stock market and give you precise rules for using oil prices to get in and out of stocks.

Key Points:

◆ Oil is essential to all we do.

◆ Since 1973, the economy and stock market have danced to oil's tune. Sharp rises in oil prices have led to recession/stagflation and plummeting stocks, while declining prices or prices that are just mildly up-trended have led to good times.

◆ For most of the 1990s, whenever oil prices rose above $20 a barrel, they came down. But starting in 1999, prices remained uptrended as the world reached the point where it was consuming all the oil that non-OPEC producers could provide.

◆ Rising oil prices will likely lead to an inflationary economy punctuated by occasional deflationary scares.

CHAPTER 2

Our Amazing
Oil Indicator

We've delineated in broad strokes how since the early 1970s rising and falling oil prices corresponded to periods of weakness and strength in the economy and financial markets. Here we'll zoom in on the stock market and show you how to use oil prices on an ongoing basis to ensure that you are in the right stocks at the right time. It will become clear just why we said earlier that oil would be our "desert island, one phone call" indicator of choice.

The basic underlying relationship to grasp is that abrupt and steep rises in oil prices are hard for the economy to absorb and hence bad for stocks. By contrast, falling oil prices, or even prices that rise but do so fairly gradually, permit the economy and stocks to flourish. Why is this so? Well, imagine you found out your salary was being cut. If it was just a small cut, you might grumble but you could adapt—you wouldn't have to make major changes in your lifestyle. If it were slashed a lot, though, it would be a different story. That idea is at the heart of our indicator. The rest is details—determining exactly how to define abrupt rises and gradual rises so as to formulate the most useful investment guide.

Oil Moves and Stocks

First, let's look in more detail at how stocks have responded to various moves in oil prices. Figure 2a, "Oil Prices and the S&P 500," depicts oil prices and the stock market since 1973, summarizing the maximum losses and maximum gains in the market—as measured by the S&P 500—within eighteen-month periods following moves in oil prices. These results give a snapshot of how much risk you face when you buy stocks after various moves in oil.

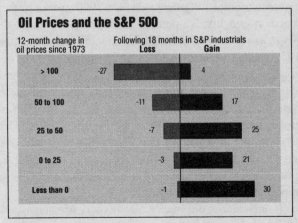

Oil Prices and the S&P 500

| 12-month change in oil prices since 1973 | Following 18 months in S&P industrials |
	Loss	Gain
> 100	-27	4
50 to 100	-11	17
25 to 50	-7	25
0 to 25	-3	21
Less than 0	-1	30

Figure 2a

Here's how we constructed the table. We looked at oil prices from the start of 1973 through the first quarter of 2003, noting where they were at the end of each month during those years and then comparing that price to their price twelve months earlier. Then we looked at where the S&P 500 index stood at the end of each month compared to where it stood one month later, two months later, and so on, for eighteen consecutive months.

The results are staggering. When oil rose by 100 percent or more over a twelve-month period, stocks during the next eighteen months experienced an average maximum monthly decline of 27 percent. (To better understand what "average maximum decline" means, suppose there were just two months in which oil prices were more than twice as high as they were twelve months earlier. For example, suppose in February 2000, oil prices were twice as high as they had been in February 1999, and in June 1974, prices were double their June 1973 level. In the eighteen months following February 2000, stocks in their worst month lost 34 percent of their value, while the worst loss in any of the eighteen months following June 1973 was 26 percent. Add 34 to 26, divide by two, and you get an average maximum loss of 30 percent. As it happens, there were fifteen months in which oil prices were more than double their year-ago levels. Thus the average maximum comes from averaging fifteen worst losses rather than just the two.) Note that we're not talking about what stocks did from the start of the eighteen-month period to the end—we're looking at their behavior at periods within that time span.

In no instance did stocks fail to experience at least one period of sharp decline. The smallest decline was 13 percent. True, sometimes stocks recovered by the end of the period, but never by much.

It's clear from the above that during the eighteen months following a 100 percent rise in oil, there also were times when stocks rose—it wasn't straight down. But when you look at stocks' maximum gains during the eighteen months following a 100 percent rise in oil, the average maximum gain was only 4 percent.

In other words, when oil prices doubled in a twelve-month period, if you had stayed out of the market, you risked missing, on average, a mere 4 percent gain at some point during the next year and a half. But if you bought stocks instead, you were more likely to see them shed 27 percent at some point during the next eighteen months. You stood to lose a lot and at best to gain just a little. These results are particularly compelling because they occurred in the context of a thirty-year period in which stocks were sharply uptrended. The evidence is overwhelming that no investor should buy stocks without first looking at the recent direction of oil prices.

When oil prices fell, the risk/reward ratio reversed, and dramatically so. Whenever oil prices dropped over a twelve-month period, during the next eighteen months, while there may have been periods during which stocks declined, the average maximum decline was only 1 percent. Meanwhile, the average maximum gain in stocks during that same eighteen-month period was a sparkling 30 percent. A 30:1 reward/risk ratio is almost too good to be true, but true it is. Believe in it and use it.

In thinking about our oil indicator, keep in mind that investing is never a matter of certainties—it's always a matter of probabilities. Stocks are inherently volatile, and even the strongest markets don't go straight up, or the weakest markets straight down. Our oil indicator can't guarantee results in a particular week or month—no indicator can. But it does tell you what your maximum gain or loss is likely to be over the following eighteen months.

Look at it like this. Suppose you were planning an eighteen-month cruise around the world, and before leaving you buy a stake in the S&P 500, for example, by buying the Vanguard 500 Index Fund. While you're away you plan to make occasional calls to your broker just to see how your stocks are doing. The numbers in figure 2a give you an indication of the most you are likely to be up or down anytime you call—or to put it differently, of the most thrilling call or the most depressing call you are likely to make. Clearly, you would want to start your cruise with the year-over-year change in oil prices as low as possible.

Doubling Your Gains

For an investor, as well as for investment advisers, the question is how to draw upon these compelling relationships to devise the most useful guide to buying and selling stocks. To some extent this is arbitrary. If history holds true, virtually any rule based on oil prices will sharply improve your results compared with how you'd do with a simple buy-and-hold approach. For instance, suppose that between mid-1973 and mid-2003 you had decided to sell stocks anytime oil prices rose by as much as 50 percent over a twelve-month period. You would buy if oil prices rose by no more than 25 percent, a point at which risks and rewards are much more favorable.

Following these two simple rules, you would have avoided the 1973–74 bear market, much of the turbulence of the early 1990s, the market crash of 1987, and some of the short bear market in 1990. You also would have side-stepped much of the massive decline that occurred at the start of this decade.

Now let's look at a slightly different rule and see how it would have worked out in terms of specific stock market gains. Suppose, for instance, that during the same thirty-year period you got out of stocks whenever the year-over-year rise in oil prices was 80 percent or greater. You got back into stocks whenever the year-over-year change fell to 20 percent or less.

In evaluating such rules, you typically use a buy-and-hold strategy as your benchmark. We'll make this comparison easy: we'll compare buying and holding the S&P 500 during those thirty years with buying the S&P 500 only when the signal from oil was favorable and selling whenever the signal was unfavorable. When you sell, we'll assume you put your money into T-bills or some equivalent short-term money market instrument.

Figures 2b, "Following Oil vs. Holding the S&P 500," and 2c, "Oil's Buy and Sell Signals," together sum up the results. As they show, during the months our oil indicator would have kept you out of the market, the S&P 500's compounded change was minus 25 percent. During those same months, cash returned nearly 40 percent. This is a massive differential. It is so massive that even though your time on the sidelines was less than four years out of thirty, you would have multiplied your original investment seventy-fold by using the oil indicator, compared to multi-plying it about thirty-five-fold with a buy-and-hold strategy.

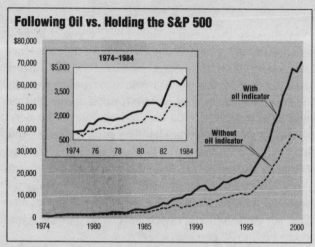

Following Oil vs. Holding the S&P 500

1974–1984

With oil indicator

Without oil indicator

Figure 2b

Oil's Buy and Sell Signals

Advice	Month/Year	Total return T-bills	Total return S&P 500
Sell	Jan. 1974	6.7	-26.0
Buy	Dec. 1974		112.0
Sell	Feb. 1980	24.0	18.3
Buy	Dec. 1981		231.0
Sell	Jul. 1987	1.8	-21.0
Buy	Dec. 1967		36.5
Sell	Sep. 1990	2.9	14.0
Buy	Jan. 1991		397.0
Sell	Nov. 1999	6.5	-4.0
Buy	Dec. 2000		

	While negative	25%	Loss
	While positive	6900%	Gain
	Total return S&P 500 during period	3400%	

Figure 2c

This is such a significant difference that we want to make sure you fully appreciate it. To put it another way, an investor who put $1,000 in the S&P 500 in 1973 and left it there for thirty years would have ended up with $35,000. Not bad. But an investor who started with that same $1,000 and followed our oil indicator would have nearly $70,000 thirty years later, an extra $35,000. And that's if you start with just $1,000. If you had a starting stake of $20,000, say, and followed our indicator, you'd have wound up with an extra $700,000 in gains compared to the investor who bought and held. And so on. The differences are enormous, and they grow more significant the more you are able to invest.

Using the Oil Indicator in the Years Ahead

Playing around with various trading rules as we did above was basically an exercise designed to demonstrate how effective it is to use oil prices as a guide to the market and how doing so would have benefited an investor in the past. All well and good. But how will we be using oil prices in the years ahead? What levels of change should constitute a signal, and when we get a signal, what exactly should investors do? For that matter, where do you go to get month-end oil prices? We answer these questions in detail below.

Check Your Oil: To construct our indicator, we began by obtaining month-end oil prices for every month going back to January 1973 and continuing to update them at the end of every month. Month-end oil prices are available online, at the Web site http://www.tax.state.ak.us/prices/.

Once there, the easiest option to follow is to click on West Texas Intermediate under monthly oil prices.

An alternative way to get year-over-year changes in oil prices is to check the back pages of the *Economist*, which carries these figures in each issue. Or any broker with access to a Bloomberg machine can get you the information in an instant.

Next we entered the prices into a computer spreadsheet and calculated the year-over-year change in prices. (Many of you will simply take us at our word when we tell you what we found, but if you're skeptical and want to check our work, simply writing the prices down on a piece of paper and using a four-function calculator will get the job done.) We did this for every reading. This gave us the raw information we needed.

Red Light, Green Light: Next we had to decide what particular level of change in oil prices constitutes a "signal"—that is, tells us it's time to take action. As we noted earlier, this is somewhat arbitrary. We eventually settled on an 80 percent year-over-year rise in oil prices as the level constituting a "negative" (deflationary) signal, meaning we should get out of most stock positions. When the rise in oil prices drops to 20 percent year over year, it constitutes a "positive" (inflationary) signal, freeing us to jump back into the market with confidence. There is nothing magic about these numbers; these are simply the levels that based on historical standards suggest that in one instance—20 percent—a rise in prices will be gradual enough that the economy can tolerate it, while in the other—80 percent—the rise will be so rapid that it will create uncertainty and discombobulation. But other levels, say, 25 percent and 75 percent, might work just about as well. It's the principle that counts.

Investment Road Map: But what do you do when you get a signal? In the discussion a few paragraphs above, where we contrasted a buy-and-hold strategy with an oil indicator strategy and saw our return double, the results were based on buying and selling "the market." And this was not in any way a mere theoretical exercise: it is entirely possible to buy and sell "the market" by investing in index funds that include all the stocks in particular market indices. In our comparisons, we assumed you were buying and selling the Vanguard 500 Index Fund, which is a no-load fund that can be purchased through the Vanguard Group.

But while index funds are a convenient way to showcase our oil indicator's performance, they won't be good investments in the years ahead, and when oil flashes a positive signal, we don't suggest that you buy "the market." That's because, as we explain later, we believe that oil prices have embarked upon a long-term uptrend. It is likely to be interrupted from time to time, but the overall direction will be up. Buttressed by other trends, this will lead to an economy that favors energy stocks and other investments leveraged to inflation (but that nonetheless will remain vulnerable to deflationary scares).

Under these particular macroeconomic conditions, investors should ignore the broad middle of stocks that are represented in the major averages. Instead, *the key to using our oil indicator for maximum gains will be to focus on an alternating mix of inflation and deflation positions*.

When oil prices rise too sharply—when our oil indicator flashes a negative signal—the economy is likely to get the deflationary jitters, meaning it is in danger of contracting, and investors should shift more heavily into risk-reducing investments that we refer to as "deflation hedges." We describe these in detail in chapter 17: they include bonds, zero coupon bonds, and short-term money instruments. When oil prices are rising at a more moderate pace—when our oil indicator is positive—it signals that the economy's predominantly inflationary bent will be able to reassert itself, and investors should emphasize inflation hedges, described in most of the chapters in part II. Following this approach will be the key to using our oil indicator to maximize your gains in the years ahead.

More on Our Indicator

You might have a few questions about the logic behind our indicator. For example, you might wonder why once our indicator flashes a negative, or deflationary, reading, meaning oil prices are 80 percent or more higher than a year earlier and it's time to abandon most of your stock positions, you have to wait until you get a reading of 20 percent before switching back into inflationary plays. After all, when the indicator is in positive, or inflationary, territory, you might get a reading of 40 percent or 50 percent or more and you don't switch into deflation hedges. And yet once you've hit deflationary territory, a reading of 40 percent, 50 percent, and so on is not good enough to signal that deflation has been vanquished. Why the discrepancy?

Perhaps the best way to understand this is by way of an analogy. Think of our oil indicator as a thermometer. When the year-over-year change in oil prices is 20 percent or less, it is equivalent to a temperature of 98.6 or slightly below.

A reading of 80 percent, by contrast, is comparable to a temperature of 105 or higher. If you have a fever that high, you're really sick—so sick that your very survival is at stake. You'd likely be hospitalized and vigilantly watched over. No one would take any chances with you, or risk releasing you too soon—the consequences would be too severe.

But what does it mean when your temperature is somewhere between 98.6 and 105? It depends in large measure on where it was before. If you had a temperature of 105 and it fell to 103, you and those taking care of you might be encouraged, but you still would be considered a very sick individual, and the fear of relapse would be ever present. Not until your temperature fell all the way back to normal would you be pronounced cured.

On the other hand, if your temperature had been normal and then rose somewhat, perhaps to 102 or 103, you wouldn't panic. You'd be concerned, and you'd certainly feel under the weather, but you'd be pretty confident that whatever ailed you was something you could deal with, an ordinary flu perhaps, something that would soon abate. And most of the time you'd be right. Only if your temperature continued to rise would you start to get really alarmed.

And so it is with our oil indicator. An 80 percent year-over-year rise in oil is major, and it indicates that the body economic runs a grave risk of becoming truly diseased. And once the economy becomes that sick, you have to fear that it won't recover so quickly, because such an abrupt rise in oil prices tends to wreak long-lasting economic havoc. You need to hang tight until you know for sure that the economy is on safe ground. A reading of 50 percent or 40 percent doesn't offer certain enough assurance. That's why you want to wait for a reading of 20 percent or below before finally shifting from deflationary to inflationary positions.

Oil is a remarkable stock market indicator. Perhaps, though, you're still not convinced. Perhaps you're thinking that other more commonly followed indicators, such as corporate earnings or interest rates, would work just as well. If so, you'd be wrong. Take corporate earnings, for example. In 1986, earnings slumped, yet stocks soared; similarly, in the early 1990s, stocks rallied despite stagnant earnings. And in 1987, when profits were rising sharply, stocks were hammered. The same was true in 2000: profits were still in a steep uptrend when stocks began one of the most vicious bear markets ever. If you go back in time, in 1973–74 profits were rising even as stocks were falling apart.

As for interest rates, the conventional view is that rising rates are bearish for stocks and falling rates are bullish. One basic reason for this assumption is that as rates rise, bonds offer higher yields and thus seemingly become a more attractive alternative to stocks. This view is right except when it's wrong, and unfortunately, it tends to be wrong a lot. In 1996 bond yields rose and so did stocks—strongly. In November 1982 the great bull market of the 1980s was just getting under way. Bond yields were under 11 percent. By early 1984 they had climbed to 12 percent. Over that same period of time, stocks soared by about 20 percent. More recently, in the 2000s a steady decline in rates failed to check a massive bear market. These are not exceptions that prove the rule, because there is no rule. The only rule is "watch oil."

Why Oil Matters

It's worth pausing to consider a point that might seem somewhat obvious but that is important to understand: How did oil get to be so essential in our lives? Why does it play such a starring role? What's the critical connection between oil on the one hand and our economy and stock market on the other?

Oil, first of all, is an utterly vital commodity, the energy source that is easiest to transport, store, and use. In a typical day millions of barrels of oil are shipped around the world. Hundreds of millions more barrels are in storage, from where they can be retrieved with relative ease and sent where they are needed. Refineries chemically convert oil into various products, including distillates and gasolines. Distillates are used primarily to heat homes and factories. Gasolines, of course, are used to power internal combustion engines. Oil also is used as a feedstock in the chemical industries. Plastics, for example, are made from oil.

Clearly we cannot run our economy without oil. Day after day, all of us rely on oil, in more ways than we typically think about. It's not just when we drive our cars or turn up the thermostat in our homes—virtually everything we touch or consume, from paper towels to English muffins to the furniture we sit on, was transported from someplace else, requiring oil. And, of course, as the movie *The Graduate* so memorably stated, in today's world plastics are indispensable.

As figure 2d, "Our Ongoing Oil Habit," shows, oil currently accounts for nearly 40 percent of total energy use in the U.S., which makes it the largest contributor to our energy needs. And as you also can see, that percentage has held more or less steady since the end of World War II. In and of itself that statistic says a lot about how essential oil is. Even though oil is the only energy source where we are not largely self-sufficient, we have done little to reduce our relative dependence on the stuff.

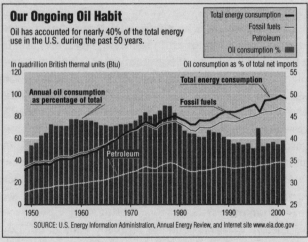

Figure 2d

While we use less oil per unit of growth today than pre-OPEC, we still use more oil today than we ever have. Indeed, by 2003 we were using 15 percent more oil than in 1986, when oil prices touched lows close to $10 a barrel. Clearly, oil is both a critical commodity to our econ-

omy and of continually growing importance. Any way
you look at it, it is more critical to us than any other commodity. We cannot get by without oil, we can't grow our
economy without oil, and moreover, if you extrapolate
from recent history, we can't grow our economy without
ever-increasing amounts of oil.

Another way of looking at oil and the tremendous
leverage it holds over our economy is to realize that there
is nothing good about rising oil prices and almost nothing
bad about falling oil prices. Accepting that oil is vital to
our economy, and keeping in mind that the ideal economy is one that grows without inflation, one way to view
rising oil prices is as a tax increase that has no redeeming
characteristics. When the government actually raises
taxes, it may have a depressing effect on the economy but
can be seen as a long-term positive if it reduces government deficits. If the government borrows less because it
has raised taxes, interest rates can decline and the economy can benefit from the lower cost of money. Moreover, it's possible to argue that the lower cost of money
also spurs investment, which can lower inflation. The
Clinton economic plan, which included higher taxes, is
widely credited with helping to bring on the prosperity
that characterized most of the 1990s.

By contrast, higher oil prices slow the economy without adding to government coffers. Not only does expensive oil lead to slower growth, but because oil is so
pervasive throughout the economy it raises inflation,
which can lead to higher interest rates and slow the economy further. To be absolutely fair, higher oil prices can
confer some benefits. Oil companies' profits may rise.
But in the U.S., since most of our oil comes from other

countries, this benefit is very minor—especially since oil production is declining here. In the pre-1970 period, when oil production was still rising, higher oil prices would lead to increases in production, more jobs, and ancillary benefits. Today, at best higher prices might mean increased profits for oil companies, who would use the bulk of the increase to explore for oil and gas outside the U.S.—not much of a trade-off for slower growth and higher inflation.

Just as high and rising oil prices are an across-the-board negative, low and falling oil prices are an all-out positive for the economy. Low oil prices are equivalent to a tax cut that instead of depleting government coffers helps build them up. When the government cuts taxes, it may lead to higher inflation by goosing the economy too much. Higher inflation may lead to higher interest rates, which slow down the economy. In sum, you go nowhere fast. When oil prices come down, however, you get lower inflation right off the bat, and it can be a double-barreled plus for growth in that lower inflation reduces interest rates, further stimulating both consumer and business spending. No wonder that the market has always climbed when oil prices retreat.

There is another dimension to the oil/economy equation. Americans, we believe, like to think of themselves as self-sufficient, beholden to no one, and able to take on the world at any time, any place. Yet when it comes to one of the things we need most, oil, we are anything but self-sufficient, depending on other nations for more than 70 percent of what we consume. When oil prices rise, it is a sharp, unwelcome reminder of our vulnerability, and it has a psychologically dampening effect on both consumer and business spending.

As evidence, consider the different reactions to price changes in natural gas and in oil. Natural gas, which we use for heating and for electricity, is comparably vital to our economy as oil, and for many uses it is a potential substitute. Yet even stunning increases in natural gas prices rarely make the headlines—even when prices are rising much faster than oil prices. In 2000, natural gas reached levels about eight times the lows in the 1990s. Oil prices rose to a bit more than three times their lows. But oil got all the attention. Why? Because we produce all our own natural gas right here at home. Even if natural gas is in short supply, we feel we can do something about it. For oil, we have to depend on others. To put it a bit differently, high oil prices are an economic, political, and psychological problem. High natural gas prices are simply an economic problem, and one that we feel is within our control.

The Rational Investor

The above detour into why oil counts is important because it shows that there is a cogent explanation for why our oil indicator works so well. It's based on real economic relationships that are grounded in both logic and history. After all, people are always coming up with seemingly successful approaches to forecasting the market, and even the most far-fetched have their ardent followers. Following oil, by contrast, is about as logical and hardheaded as you can get.

Having guidelines that keep you rational and hardheaded will be more important than ever in the turbulent years that lie ahead. Moreover, they would have saved millions of investors tons of grief and money in the years recently gone by. The late 1990s, for sure, were a time when rationality was thrown to the winds as stock prices soared in seeming defiance of basic economic laws. In those years, analysts made all kinds of pronouncements with great abandon, and investors were eager to be sucked in. Logic went begging.

One typical example was a report in the 1990s from a Paine Webber analyst urging investors to load up on Qualcomm, maker of wireless communications products. The company's stock already had risen thirty-fold in two years, an awesome gain. No matter. According to this analyst, the company's profits would grow 30 percent a year for the next decade. And this easily justified the stock's doubling from its already exalted level.

What's amazing is that almost no one blinked twice at this report. Yet if you looked at its underlying assumptions, it clearly presumed that over the next ten years at least two cell phones would be sold for every man, woman, and child on earth. There wasn't a shred of plausibility to this assumption, but in the greed and frenzy of the great bull market of the 1990s, no one noticed or cared. Not surprisingly, the stock subsequently fell some 80 percent in the 2000–2002 crash in the Nasdaq, and plenty of investors got badly burned. If those same investors had been more skeptical, and had looked at the underlying economic realities—in particular if they had paid attention to what was going on with oil—they would have been sitting safely and no doubt smugly on the sidelines, their profits from earlier years intact.

In the late 1990s it had become part of the common

wisdom to proclaim that stocks, despite sky-high P/Es (price/earnings ratios), were fairly valued. We had, many argued, entered a true Goldilocks economy, one in which everything was just right, and the explosive gains in the stock market, far from being excessive, merely reflected this new reality. Talk about wishful thinking! Rising oil prices were arguably the pin that pricked the tech bubble. Oil is real, and by heeding its message you can inoculate yourself against infection by fantasy fever.

With all the above as background, obviously the burning question is: what's next for oil prices? In the next two chapters we look at why prices in all likelihood are headed higher, though probably with occasional dips and pauses. If we're right, it means that in coming years we're headed for a turbulent and largely inflationary economy and a rocky stock market. To outsmart such a market will require selective, disciplined, and highly knowledgeable investing. In such a tough and unforgiving investment environment, making the right choices will mean the difference between big gains and watching your capital disappear. And it all hinges on oil.

Key Points:

◆ When oil prices are sharply higher than year-earlier levels, the risks in the market over the next eighteen months overwhelm potential gains. The situation operates in reverse when oil prices are relatively stable.

◆ Specifically, our oil indicator flashes a negative signal whenever oil prices are 80 percent or more higher than a year earlier. Shift most of your assets into deflationary hedges such as T-bills and bonds.

- ◆ When the year-over-year rise in prices drops to 20 percent or less, snap up inflation plays, such as energy stocks and precious metals.
- ◆ To track oil prices visit the Web site http://www. tax.state.ak.us/prices/, or look at back issues of the *Economist*. (The authors' investment letter *The Complete Investor* tracks our oil indicator on a monthly basis.)

CHAPTER 3

The Geological
Lowdown

In this chapter we look at how much oil is left in the world and at how much of it can and/or will be developed. It may seem heavy on science. But in many ways this chapter is the absolute crux of our book, the central argument on which everything else hinges. If you read nothing else, read this chapter. In fact, if you read nothing else, read the following paragraph, because in it we summarize, as succinctly and strongly and simply as we can, why we are likely headed for an oil crisis that will—literally—be the oil crisis to end all oil crises, one that will set the parameters of our economy and financial markets for the next decade or longer.

So here goes: the U.S. and the world will continue to need growing amounts of oil to keep economic growth on track. But at some point before 2010 the world will become incapable of producing the extra amount of oil needed. How quickly we reach this point will depend almost entirely upon developments in the Middle East. In a best-case scenario we will have until the start of the next decade. That will be the case if, and only if, Saudi Arabia suddenly becomes willing to develop all its reserves *and* if its reserves are as extensive as it has said they are. Both of these are tremendous ifs. On a worst-case basis, oil supplies will become inadequate within the next two years. Technology will not come to our rescue via the sudden discovery of better ways of extracting oil from existing fields; efforts to explore and develop new fields will be of puny significance; and efforts to conserve oil will be meaningless compared with the huge spurts in usage worldwide, particularly in China. The widening gap between demand (growing) and supply (growing far less rapidly, or even shrinking) will push oil prices ever higher—up to as high as $100 a barrel before the decade is over. There is just one silver lining to this heavy oil-induced cloud: sky-high oil prices will be the essential condition that will, finally, propel us into making the historic transition to alternative, renewable, nonpolluting fuels.

As we indicated, the above paragraph pretty much says it all. From its assertions flow everything that awaits us in coming years—oil shortages and sharp rises in oil prices, inflationary pressures, economic turbulence, and a stock market that will penalize investments in financial assets and reward investments in the kinds of real assets scorned in the 1990s. And it all hinges on the esoteric world of geology.

Hubbert's Law

Starting in 1999, as we noted in chapter 1, oil prices became supply-driven as OPEC became the only real game in town. We argued that this situation will persist—it's not a temporary imbalance that will be corrected in the near future by the West getting on the stick and finding more oil. And if this conclusion seems pessimistic, it gets worse. It's not just that the non-OPEC world isn't going to be able to produce the oil the world needs to grow. OPEC producers themselves are reaching limits that will exacerbate the world's oil woes. Increasing oil supplies sufficiently to meet the world's needs for continued growth will soon be beyond anyone's abilities.

Let's look at the non-OPEC world first. Why are we so pessimistic about the possibility of non-OPEC nations developing the ability to wring greater amounts of oil from their existing fields, or discovering new fields with meaningful capacity? There are two answers: the historical evidence and, equally compelling, the work done decades back, and brilliantly verified since then, by a prescient American geologist named M. King Hubbert.

Writing in the 1950s, Hubbert predicted that production of U.S. oil fields would peak in the early 1970s. When he made this claim, oil production in the U.S. was rising steadily, seemingly inexorably, and few if any experts at the time took him seriously. But he turned out to be absolutely right, as you can see from looking at figure 3a, "Declining Domestic Oil Production." Oil production rose in the U.S. throughout the 1950s and 1960s. In 1970, however, production began to decline.

Hubbert's key insight was that production peaks once half the oil in any field has been extracted. It's a bit like eating a loaf of bread, where with each meal you eat half of what you have. The first meal you get a lot of bread. But after that you get declining amounts. It's not that

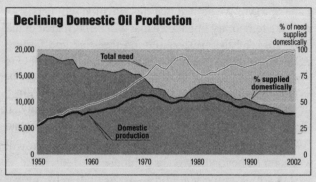

Figure 3a

you'll ever run out of bread entirely—there always will be some small crumb left to cut in half. But it won't do much to satisfy your hunger.

If this seems too simplistic, the basic idea has held up remarkably well with respect to oil production. Any oil field contains only a finite amount of oil—it is not generated anew—and even though it may be true that we never will extract all the oil a field contains, it is equally true that we will produce far more in the early stages of production than later on. Starting from this commonsense premise, Hubbert applied some fairly sophisticated curve-fitting, which led him to conclude that by the early 1970s all the low-hanging fruit would have been picked in U.S. oil fields. From that point on it would get tougher to get out the oil that remained, and as a result, production would decline. Typically, the point at which production starts to decline is when half of what was there at the start has been removed.

Hubbert's work has applied equally successfully to many deposits outside the U.S. The North Sea is a major example. It was defined as an important oil and gas deposit in 1969. At the time, Britain and Norway, the two major beneficiaries of the discovery, were struggling economically. In fact, it was fashionable in the early 1970s to refer to Britain as a burgeoning third world economy. To develop North Sea oil as rapidly as possible, the most advanced technology available was brought to bear. As a result, in thirty years the North Sea produced about 15 billion barrels of oil, and in the process Norway became rich and Britain shed its status as a third-rate economy. The problem is that there are no encores. Production in the British portion already has started to decline, and Norway is very close to peak production.

The North Sea example helps point up the limits of technology in extending the life of oil fields. Some have argued that technology will ride to the rescue, that there have to be ways of getting more oil out of older fields— after all, no one is denying that billions of barrels of oil still lie below ground. But the evidence is strong that technology won't bail us out. The U.S. is easily the most technologically advanced nation in the world, and yet it hasn't found a way to halt the decline in production over the past thirty-plus years. The only increase came with the development of Alaska's North Slope, and already production there has started to decline. And as the North Sea shows, technology is a double-edged sword. That is, by using the most advanced technology, you can get out more oil up front than you could previously—but this merely makes the day of reckoning, when production starts to decline, come sooner.

The Mystery of Mideast Reserves

What has happened to oil production in the U.S. will almost certainly happen worldwide, and much sooner than most people expect. An authoritative analysis of this looming decline in production is presented in the book *Hubbert's Peak: The Impending World Oil Shortage,* written by Princeton professor Kenneth S. Deffeyes, a former colleague of Hubbert. Deffeyes sought to apply Hubbert's geological precepts on a worldwide basis. Other geologists have made the same effort. And while they haven't all picked precisely the same year, the degree of consensus among them is quite remarkable: most concur that worldwide oil production will peak sometime before 2010. The discrepancy in dates, which is small to begin with, stems from varying estimates of Mideast reserves, and to a much lesser extent FSU (former Soviet Union) reserves.

There are two ways to estimate remaining reserves and thus the point at which production will peak. The first, which was Hubbert's method, is to analyze production curves. Production will typically follow a "normal" curve, similar to the one teachers often follow in determining grades, the same that defines the distribution of many characteristics in the population, from height to spatial abilities. One reason production follows such a predictable path is that there are reliable relationships between new discoveries and the production that flows from them. In the Middle East, however, there is a problem with this approach: there has been only scant exploration in the past generation. Thus, trying to estimate reserves based on the standard Hubbert methodology doesn't get you too far.

The second approach is to make a direct geological assessment of remaining reserves, by physically examining particular deposits and analyzing the characteristics of the land in which they lie. For countries outside OPEC, both approaches seem to work equally well, and they produce very similar results.

In the Middle East, because of the lack of exploration noted above, the only things we have to go on are geological estimates, but unfortunately, though not surprisingly, in this part of the world such estimates are highly suspect. The most thought-provoking work on Middle East reserves has been done by a geologist named Dr. Colin Campbell. In 1998, Campbell and a colleague Jean Laherrère published an article in *Scientific American* detailing numerous oddities about estimates of Middle East reserves. (Campbell's work also is featured in a Web site named Hubbertpeak.com.)

One thing Campbell and Laherrère noted about the Middle East was that even though a good deal of production was occurring, the extent of reserves reported remained amazingly constant from year to year—and then jumped dramatically. For example, between 1980 and 1989, Saudi Arabia, the country with the largest reserves, reported only nominal changes. Then, in 1990, the country reported that its reserves had grown by nearly 90 billion barrels—the equivalent of three North Seas. Wow!

A similar mystery enfolds the oil picture in Iraq and Iran. After a period in which its reported reserves had remained more or less constant, Iraq in 1988 announced that they had more than doubled, to 100 billion barrels! Equally miraculous, since then, Iraq's reserves, as officially reported, have remained at precisely 100 billion barrels despite continued production and a total absence of exploration. And Iran has followed an identical pattern, nearly doubling its estimate of reserves in 1988 after years of stagnation. In and of themselves these figures would defy belief; throw in the context of the Iran-Iraq War in the 1980s and they become even less likely—is it really possible to imagine that each of these countries carried out preternaturally successful exploration efforts while engaged in a bitter war?

All these actions by the Saudis, Iraq, and Iran—reporting reserves constant for years despite production and then announcing huge increases in reserves—strongly suggest that OPEC's reserves are overstated. And as Campbell points out, that is exactly what you'd expect given the nature of the OPEC cartel. That's because quotas for the members are determined by production capacity, and production capacity is directly related to reserves.

It seems indisputable, then, that OPEC has less oil left than it claims. The question is how much less. The answer, which may lie in the calculating abilities of some little-known geologist, will determine just when the world will hit the point when even the most heroic efforts will fail to increase oil production. There are different ways to get at an answer. But no matter what method you use, the outlook is not rosy. At best, we have until the end of the decade before the oil squeeze becomes a true threat to the capitalistic system. More likely, we have less time than that. And it's possible that in just a year or two, world oil production will peak and begin an inexorable decline that will send oil prices skyrocketing.

The Will to Drill

The extent of Middle Eastern reserves is just one part, although obviously a very important part, of the picture. The other unknown factor is whether, whatever the extent of its reserves, the Middle East has any intention of developing them. In conjunction with something more readily estimated—worldwide demand for oil in coming years—this will give us a clearer idea of when the ax will fall. And again, the outlook is exceedingly alarming.

An article in the March/April 2002 issue of *Foreign Affairs* tackles these questions—and is quite enlightening on a number of counts. Authors Edward Morse and James Richard write: "Even before September 11, concerns had been raised over American reliance on Middle East oil. Global oil demand has been increasing by between 1.5 and 2 mbd (millions of barrels per day) each year, a rate of growth with alarming long-term consequences. The U.S. Department of Energy and the International Energy Agency both project that global oil demand could grow from the current 77 mbd to 120 mbd in 20 years, driven by the United States and emerging markets of South and East Asia. The agencies assume that most of the supply required to meet this demand must come from OPEC, whose production is expected to jump from 28 mbd in 1998 to 60 mbd in 2020. Virtually all of this increase would come from the Middle East, especially Saudi Arabia." This paragraph came on page 18 of the article. Twelve pages later the authors make another statement: "Riyadh, on the other hand, might have vast known reserves, but it also has a closed state monopoly. Most alarming, Saudi Arabia has been unable for 20 years to increase its production capacity. Nor is its position unique: few OPEC countries in 2002 have more production capacity than they did in 1990 or 1980."

Look at these two paragraphs together (and the fact that the authors didn't do so is in itself quite scary, for it suggests that even the best-informed experts in this field have yet to really grasp the gravity of the situation), and the problem becomes crystal clear. Demand for oil will grow sharply. To obtain that oil we are relying on Saudi Arabia and other Middle Eastern countries—who have barely increased production since the early 1980s—to suddenly launch a massive development program, one so huge that it will approximately double current production. As icing on the cake, the only basis for believing these countries even have the reserves we are counting on them to develop is the reserve estimates they themselves have issued and that almost certainly vastly overstate what their oil endowment is.

If we know all this, the Saudis themselves surely know it as well. After all, the Saudis own shares in U.S. oil companies, and they are well aware that the costs of finding oil are rising and that companies are having an ever more difficult time replacing reserves that are used up. They know that the typical operating oil well is being depleted at the rate of about 4 percent a year. And indeed, they must know that even in 2002, when world economic growth was well below average, oil demand still rose by nearly 1.5 percent. Moreover, even if their undeniable hatred of/ambivalence toward the U.S. is a conscious or even subconscious factor, the Saudis know other countries from China to France are in the same boat and need oil for growth. In short, the Saudis are surely aware that a failure to develop their reserves will put extraordinary pressure on the entire Western world.

They also know they will be able to sell all the oil they develop. Developing their reserves, in other words, would seem to be a no-brainer for the Saudis. Not only would they attract a ton of Western money into their country to fund the necessary massive development projects, they also would ensure themselves of dramatically increased income for a long time to come.

So what gives? Why haven't the Saudis launched a crash program to develop desperately needed oil? It seems clear that the answer has nothing to do with economics and everything to do with the nature of the Saudi ruling class. Saudi Arabia is ruled by a monarchy that is desperate to maintain its way of life and hold on to its power over the Saudi masses. One thing above all threatens its iron grip over the country. And that is economic growth, which would raise the living standards of the ordinary Saudi and in all likelihood raise expectations for further improvements and for some measure of self-rule—i.e., some of the trappings and benefits of democracy. This point has been made frequently and cogently, in particular by the *New York Times* columnist Thomas Friedman, who has noted that the Saudis have used their oil revenues not to foster economic growth in their country but to keep their populace oppressed and in tow.

Friedman's arguments have been based in part on his personal observations, but there is ample hard data leading to the same conclusion. As figure 3b, "Declining Saudi GDP," shows, income per capita in Saudi Arabia has been in a steady downtrend over the past decade. This has occurred despite the fact that the country has been

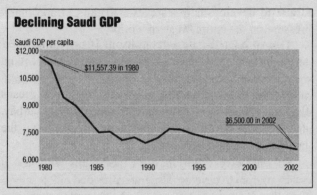

Declining Saudi GDP

Saudi GDP per capita

$11,557.39 in 1980

$6,500.00 in 2002

Figure 3b

receiving massive amounts of money—nearly $100 billion, or $4,500 per capita, a year—in oil revenues for very little effort. Given this tremendous influx of funds, it seems almost unimaginable that Saudi Arabia hasn't managed to provide a vibrant economy for its citizens, and it is evidence of repression with a capital R. Can you imagine any Western country experiencing steady declines in standards of living—especially with money pouring into the country—and no change in who's in charge? The Saudis have been dedicated to using their enormous oil wealth to protect those in power. To accomplish this, they have heavily funded their very conservative clergy and done whatever it takes to keep Western influence to a minimum.

So for the ruling Saudi family, the dilemma is that it would take Western money and Western expertise to develop the country's reserves, but anything Western is a mortal threat to the kingdom. The bottom line is that what the West desperately needs—the development of Saudi reserves—the Saudis are desperate to avoid. And avoid it they have over the past generation. If anything, their ardor has been increasing. In today's Internet-linked world, ideas from afar have a way of insinuating themselves into even the most recalcitrant society. This means the Saudis must work extra hard to keep the West at bay, and it makes them less willing than ever to permit the launching of major Western-run development projects.

Of course, it is not impossible that the Saudis will shift gears. Maybe they will have a change of heart. Under the best-case scenario, the royal family would yield to a pro-Western democratic government that would encourage development of reserves. But remember—even under this best-case outcome, we still are left with a world in which oil production is set to decline. It's just that the decline would come a little later than if the additional Saudi capacity weren't brought on line.

Challenging Hubbert's Law

To be scrupulously fair, we want to note that not every geologist accepts that Hubbert's law continues to apply to the world today or that oil production will ebb as soon as we've suggested. The arguments of the naysayers, though, don't hold up under scrutiny.

Ronald R. Charpentier, a geologist for the U.S. Geological Survey (USGS), has challenged it primarily on the grounds that it relies less on field research than on armchair analysis. This criticism doesn't acknowledge the fact that armchair analysis in the hands of a skilled analyst is always based on a massive amount of data that is then interpreted within an internally consistent framework. (It also implicitly disparages the work of such armchair scientists as Einstein and Newton.) The USGS is a lone voice in that it predicts that oil production won't peak until 2020. This is in contrast to the other major agency with comparable resources, the International Energy Agency (IEA), which serves the Organization for Economic Cooperation and Development (OECD), representing the major economies of the world. It projects that oil production will peak in just a few years at around 80 million barrels a day, compared with current production levels of 77 million barrels a day.

A second and more interesting line of challenge to Hubbert is a variant of the technology-will-solve-everything argument. It has been most forcefully stated by Michael Toman and Joel Darmstadter, when they were senior fellows at the Washington, D.C.–based Resources for the Future. In a letter published in 1998 in *Science* magazine, they argued that changes in energy prices and technology, by in effect raising estimates of the amount of recoverable oil in the ground, make it impossible to apply Hubbert's curve analysis with any accuracy. In other words, if prices go up enough, more of the absolute amount of oil in the ground will be seen by producers as reserves that are worth developing. This is true, but it doesn't affect the basic insight that after a certain period and rate of depletion of whatever absolute reserves exist, production will decline. And what's really interesting with respect to this argument is that Hubbert's original projections of when oil production would peak in the U.S. remained on track even as prices rose from below $3 a barrel in 1970 to $30 a barrel.

THE GEOLOGICAL LOWDOWN • 57

No geologist would deny that much higher oil prices may result in the addition of reserves. But there is no evidence that this will have a significant impact on world supplies. For investors, the critical point is that even under the assumptions of the most thoughtful of the anti-Hubbert contingent, the world still faces an inexorable and steep rise in oil prices.

◆ ◆ ◆

Oil has been essential to our economy for a long time. Despite its obvious problems, from pollution to its concentration in an often hostile part of the world, it has long been the fuel of choice, particularly when it comes to running cars and trains and planes, because of its convenience and the availability that has kept it relatively cheap. But nothing lasts forever. In the past, when oil supplies have tightened, it has been a question of political conflicts and deliberate cutbacks. Going forward, however, the supply constraints will result from irreducible and irreversible geological realities. This time around it's not a matter of crying wolf. The end of oil's hegemony is within sight. The repercussions will be enormous and will affect all of us.

Key Points:

- ◆ We're nearing a worldwide peak in oil production. There still will be plenty of oil in the ground, but it will become increasingly difficult and expensive to keep pumping it out.
- ◆ A key reason: Hubbert's law. M. King Hubbert, a geologist, predicted in the 1950s that oil production in a

field peaks once half its oil has been extracted. So far he has been remarkably prescient.

◆ Mideast reserves are likely less than has been stated, and there are serious questions as to whether Saudi Arabia even wants to fully develop its reserves.

◆ As geological limits lead to peaking oil production, inflationary pressures will build.

CHAPTER 4

Oil Prices: Up, Up, and Away

On February 20, 2003, the *New York Times* ran an article about the sudden jump in prices at the gasoline pump that had occurred in the preceding week or so in many parts of the country. Gas prices had risen an average of 29 cents a gallon since late 2002, and many drivers suddenly were having to shell out $2 or more a gallon. People were feeling the pinch. One couple interviewed for the piece, residents of a Seattle suburb, said that with gas prices so high, they had stopped taking separate cars to work. Instead they were keeping their Chevy pickup truck, which got only 12 miles to the gallon, at home, and dropping each other off in their more fuel-efficient Toyota.

Indeed, the rise in gasoline prices—attributed variably to the war then pending against Iraq, the strike in Venezuela, and the greed of oil companies capitalizing on the geopolitical situation—was enough to start to make a dent in a lot of people's household budgets. But despite any fluctuations in gas prices since then, the chances are good that it won't be very long before Americans look back on the days of $2 gasoline with tremendous nostalgia.

The world is fast approaching a massive clash between demand for oil and our ability to meet that demand. Economics 101 teaches that when demand exceeds supply, prices go up. And there is no question that oil prices will rise—dramatically—over the next decade. They will continue to rise until we develop viable energy alternatives and commit to their widespread use.

Rising oil prices will have several effects. First, they most likely will spur overall inflation, which in turn will increase uncertainty and make financial markets less stable. Just how unstable they become, however, will depend more on how rapidly and abruptly oil prices rise than on the absolute level they reach. Remember, the basis of our oil indicator is that economies generally can absorb relatively gradual rises in oil prices without being thrown into disarray. It's sharp and abrupt rises that throw the economy, and the financial markets, for a loop.

Second, as the example of the Seattle couple shows, as oil prices rise people will do what they can to cut back on consumption. Over the short term, such increases in conservation may help slow down the rise in prices. Ultimately, however, the imbalances between supply and demand will overwhelm efforts to conserve. In the long run, the only thing that will resolve the crisis will be a definitive switch to alternative energies.

We have no way of knowing whether oil prices will rise abruptly or more gradually. We can, however, take a stab at estimating just how high oil will go. And as we said, in the not too distant future, $2 gas will sound like an incredible deal.

Below, very briefly, we crunch some numbers so as to give you some notion of just how high oil is headed. One reason we think it worthwhile to go through this exercise is simply that it has a certain shock value. Americans have long tended to regard cheap oil as an inalienable right. When prices have risen, it's been seen as an aberration; when prices have come down, we assume the problem has gone away. Maybe it's because by nature we are an optimistic people. Or maybe it's because we are near-sighted and complacent. We assume that if we have something today, we can count on keeping it tomorrow and forever. But applying that attitude to oil has been a major mistake. It is a mistake that will shape our lives for many years to come.

Indeed, along with helping investors cope success-fully with the rocky markets that lie ahead, one of our goals in writing this book is to puncture the complacency that even today remains so prevalent, out of the belief that the less complacent we are as a country, the more likely that we will make the right choices. We feel a bit like Paul Revere, except that instead of crying out "the British are coming," we're crying out "higher oil is coming." Also, we hope that by coming up with concrete numbers, we might make you, as investors, pay even closer attention than you otherwise might do. And this could be, if not a lifesaver, at least a portfolio saver. .

Triple-Digit Oil

In trying to work out where oil is headed, it's reason-able to start with the expectation that, given the supply crunch we foresee, real oil prices—prices after inflation—

will make significant new highs. If you look at figure 4a, "Real Oil Prices, 1949–2002," you'll see that in 2002 with oil near $30 a barrel, prices in real terms were well under half of where they were in 1981. This means that just to return to 1981 levels, oil would have to jump to over $60 a barrel.

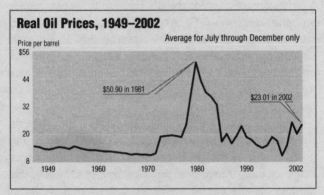

Real Oil Prices, 1949–2002

Price per barrel

Average for July through December only

$50.90 in 1981

$23.01 in 2002

Figure 4a

But we don't expect oil to make new real highs overnight, for that's not how markets work. Suppose it takes a decade for these highs in oil to be made, and that overall inflation during that decade averages just 4 percent a year. Then for oil prices in real terms to be at highs at the end of the decade, oil would be priced at above $100 a barrel. If average inflation is higher, which is what we expect, oil would climb even more. And if oil were to top its real highs by any significant margin, you are talking about higher prices still.

Any way you slice it, oil seems headed for triple-digit

levels by the end of the decade. Oil at $100 will be a minimum. Just to put this in perspective, this could mean gas prices at the pump approaching $10 a gallon.

It's important to realize, however, that this rise won't be in a straight line. Almost certainly there will be times when oil prices retreat. It's essential that you not be fooled by these periods of weakness—that they don't lead you to conclude that oil's uptrend has been permanently quashed.

There are two reasons why the uptrend will be interrupted from time to time. The first will come from the inevitable periods of slow economic growth, while the second will result from efforts at conservation. Just keep in mind that the effects of both will be transitory and won't check the rise in oil prices for long.

History offers some interesting insights here. In the 1979–83 period, strenuous efforts to conserve energy combined with two recessions enabled the U.S. to reduce its energy consumption and cut back on its use of oil and other fossil fuels. What was true in the U.S. was true to a lesser extent for the world as a whole, which also managed to curtail its energy consumption, but only through 1982.

What's really telling, though, is to look at what led to that brief period of worldwide conservation. It took a more than tenfold increase in oil prices over the preceding seven years and two U.S. recessions in 1980–82 to do the trick. It seems that if you want to believe that conservation will play a role in moderating the uptrend in the prices of oil and other fossil fuels, you have to accept that a lot of turmoil lies ahead and that oil prices will likely carry past $100 a barrel, at the least.

The China Syndrome

What's also true, though, is that today's world is very different from the early 1980s, in ways that suggest that conservation on a worldwide basis—and remember, it's worldwide demand for oil that counts—is not going to be able to do much to check oil prices. In the early 1980s, the chief engine of world growth was the U.S., followed by Europe and Japan. These economies had one important thing in common: they all were making the transition from being driven primarily by manufacturing to being driven primarily by services. Today, while Japan has sputtered a bit, the U.S. and Europe still are pulling the world's economic train—along with one newcomer. That brash newcomer is China. And China's emergence adds a whole new spin.

Depending on how you measure economic output, China is the third-, the fifth-, or the sixth-largest economy in the world. But no matter how you measure it, China is the fastest-growing major economy in the world. This means that for a while to come China is likely to be the world's leading economic engine, even more important than the U.S. China's economic heft comes from the sheer size of its population, in contrast to the U.S., where it is a matter of population in conjunction with an elevated standard of living, or income per capita. On a per capita basis, income in China is modest by world standards. Most important for our argument, per capita oil consumption in China is currently just one-half or so of the world's average. And it's about one-tenth that of many modern industrialized economies, such as the U.S.

Like other major world economies, China also is making a significant transition. But in China's case, it's not a transition from manufacturing to services, but from agriculture to manufacturing. And this means that per capita growth in oil usage is almost certain to grow rapidly. As manufacturing becomes more important to an economy and agriculture becomes more centralized, the use of oil for transportation and other purposes rises sharply. By the first several years of the twenty-first century China was using about half the energy the U.S. does. If China were just to match the world's average per capita energy use, it would double its oil consumption and use as much oil as the U.S. does today. If China continues to grow and continues to industrialize, it will end up using far more oil than the U.S.

As the experience of the early 2000s shows, China's growth is largely independent of what goes on in the rest of the world. In 2002, for instance, while most major world economies slumped, China's economy grew at a rate utterly unobtainable by developed economies—over 7 percent. And not surprisingly, despite slow worldwide growth in 2002, oil consumption for the world rose by about 1.5 percent. This is a rate of increase more consistent with rapid than with slow worldwide growth. China, obviously, was the difference.

In a later chapter we discuss why we think that efforts to mandate conservation, for example, by requiring more fuel-efficient cars, may actually backfire. Here, our point is that conservation is simply not going to do the trick—not with China easily blowing past whatever we might save, not with worldwide economic growth a necessity.

ANWR No Answer

Whatever Congress ultimately decides to do about the contentious issue of drilling in the Arctic National Wildlife Refuge (ANWR), and right now it looks as though drilling will be nixed, the issue is worth looking at, if for no other reason than it shows just how delusional it is to think we might solve our energy problems by finding new areas to drill. The area contains an estimated 10 billion barrels of oil. It might even have 15 billion. Whatever—it is still far less than the North Sea contained. It is further estimated that peak production will occur about fifteen years after development starts and that getting to that starting point, i.e., developing the reserves, will take about five years. That means that we are looking at twenty years down the road before ANWR delivers its full promise, assuming, of course, that Congress approves opening it up to development.

And what will that peak production be? About half a million barrels a day. This is a fraction of 1 percent of what the world will likely be consuming by then—it's even well under 1 percent of what the world is consuming now. Half a million barrels a day is even less than the annual rate at which wells in the U.S. alone are being depleted. We would need many ANWRs to hope to come even close to making a dent in supplying the world's future oil needs. And this analysis, of course, is totally apart from all the environmental objections to drilling in this ecologically sensitive area.

As for hopes that there may be other big deposits in the world, some new North Seas that somehow have eluded detection up to now, the odds against that are exceedingly high. Extensive exploration has turned up no such thing. Given geologists' sophisticated understanding of the characteristics that would indicate a major oil find, it isn't likely that any oil-bearing area large enough to be significant has eluded attention. There is no escaping it: absent worldwide depression, the world is headed for triple-digit oil. Investors and consumers alike should be girding themselves to deal with oil prices of $100 a barrel and higher. A new world is coming.

Key Points:

♦ Oil prices are set to soar. The only question is whether they will do so rapidly and abruptly, sending the economy and financial markets into a deflationary tailspin, or more gradually, triggering high and rising inflation.

♦ Prices will likely reach at least $100 a barrel before the decade is out.

♦ As prices rise, efforts to conserve may temporarily moderate the uptrend. Note, though, that in the 1980s it took a more than tenfold increase in oil prices to spur serious efforts at conservation.

♦ In any case, China will be a big consumer of oil, helping to keep worldwide demand strong and prices high.

♦ Oil from new exploration, including any efforts to open up the Arctic National Wildlife Refuge, will barely make a dent in our growing need for energy.

CHAPTER 5

The Debt Burden

Polonius, the father of Ophelia in Shakespeare's *Hamlet,* is generally regarded as a pompous old fool who probably deserved being stabbed in the gut by his daughter's somewhat edgy on-again, off-again suitor. But if more Americans in the 1990s had followed his advice — his famous exhortation to "neither a borrower nor a lender be" — the world would have had a better shot at warding off sky-high oil prices, and inflation, in the early twenty-first century.

Here's the connection: as oil supplies peak, the only thing that could keep oil prices from rising sharply would be declining demand. But demand for oil will drop only if there is a dramatic worldwide slowdown in economic growth. A few decades back it was feasible for inflation fighters — remember Paul Volcker? — to take vigorous steps to tamp down economic growth. Today, however, such action would run an unacceptably high risk of triggering total economic mayhem. And — here is the Polonius tie-in — the reason for this increased risk is a huge burden of debt, in particular consumer debt, that overhangs the economy.

Debt has made the economy far more fragile, making continued economic growth a necessity. Economic growth requires oil. Hence, demand for oil will remain strong. As supplies falter, there will be no way to ward off higher oil prices.

The Rise in Consumer Debt

Just how high is debt today? It's at near-record levels. As figure 5a, "Debt as Percentage of GDP," shows, by 2002, overall debt was 2.8 times the gross domestic product (GDP)—a level not seen since the Great Depression. In 1981, by contrast, overall debt was 1.7 times GDP—a lower level that gave Federal Reserve chairman Volcker the leeway to sharply raise interest rates and tighten the money supply.

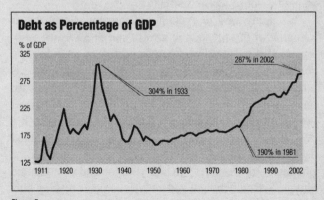

Figure 5a

Since the early 1980s, all categories of debt—government, business, and consumer—have grown faster than the economy. But consumer debt has been the fastest-growing segment.

A major reason was the coming of age of the great bulk of the baby boom generation. In the early 1980s, baby boomers in droves were starting to marry, settle down, and have families. Typically that is the stage of life at which you rack up the most debt—you need a whole bunch of things, such as a car and house, and you haven't accumulated the savings necessary to pay for them without borrowing. And the data suggest that baby boomers, perhaps inherently optimistic and imbued with a sense of entitlement to the good life, were particularly willing to go out on a limb in order to acquire homes. In 1982, the price of a typical home was about 4.2 times after-tax income. That is, if your after-tax income were $50,000, you would be willing to pay $210,000 for a house. By the end of the 1990s, the figure was 4.8—you wanted a house worth $240,000.

It's pretty easy to explain why consumer debt that is rising faster than economic growth can be a serious problem. Let's look at it from the perspective of an individual consumer first. Suppose you borrow money for various purchases. If you are earning enough money to meet your ongoing needs, make your interest payments, and pay off your debt on a reasonable schedule, the debt is no big deal. If your income rises, you can even take on additional debt. But if you take on more debt while your income stagnates—or worse, if you lose your job and your income plunges—then debt will suddenly become a crushing burden. In other words, debt makes you very vulnerable to any slowdown in income. The bigger the debt in relation to your income, the greater your vulnerability.

This is equally true for the economy as a whole. When debt is rising faster than the economy is growing, as it has over the past twenty years, such vulnerability exists writ large.

When the economy is loaded with debt, it's essential that the flow of money continues to grow. Big debt in the context of a growing economy is manageable for society at large, just as it is for an individual. What's not manageable is big debt in the context of stagnant or declining economic growth. Or, what amounts to the same thing, growing debt in the context of an economy that is growing more slowly.

The 1990s were an era of great prosperity for many. Even then, though, debt that was growing faster than incomes already was taking a tremendous toll. The evidence: a surge in consumer bankruptcy filings, as depicted in figure 5b, "Soaring Bankruptcy Petitions." In 1990, about 700,000 consumers filed for bankruptcy. By 2000

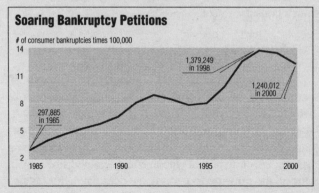

Figure 5b

the figure had soared to over 1.2 million. During the slow-growth 2000–2002 period, bankruptcy filings further accelerated to more than 1.5 million a year. Blame this trend on debt levels that grew much faster than incomes. It shows just why policymakers will be under constant pressure to keep growth strong even if it means inflation.

A related point has to do with the level of interest rates. Your debt burden, after all, is determined not just by how much debt you owe but also by the interest rate you are paying on that debt. If you owe $50,000, and are paying 1 percent yearly interest, your yearly debt payments are only $500 plus whatever principle you want to pay off. But if interest rates are 10 percent, your debt burden is far greater. Keep in mind that nominal interest rates rise as inflation goes up. As rates rise, policymakers, to prevent debt from overwhelming consumers, will be pressed to keep economic growth strong to ensure that wages rise faster than rates. This will be a tremendous push toward further inflation. In essence, inflation will beget further inflation as policymakers find their hands tied by the existence of so much consumer debt.

Home Sweet Home

A major portion of total consumer debt—about 70 percent—is in the form of home mortgages, whose growth is displayed in figure 5c, "Mortgage Debt Outstanding." According to the Federal Reserve, nonmortgage debt (credit cards, bank loans, etc.) payment as a percentage of disposable income has declined since 1980. During the same period, mortgage debt payments, as figure 5c shows, have risen by over 40 percent.

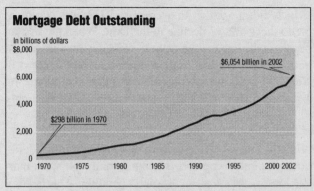

Mortgage Debt Outstanding

In billions of dollars

$6,054 billion in 2002

$298 billion in 1970

Figure 5c

There's a good reason why homes account for so much of overall consumer debt—banks are more willing to lend money for a home than for almost anything else. That's because a home is a tangible and enduring piece of property, one, in fact, whose value is likely to increase over time. This explains why consumers use their homes as collateral to get other loans, taking out second mortgages to finance various purchases or for home improvement. A huge proportion of consumer debt is secured by the home—about 72 percent. It is hard to overstate how important homes are in consumer finances.

Homes were never more critical to consumers and the economy than in 2000 in the aftermath of the tech bubble. The bear market of 2000–2002 was by any measure one of the worst in the past hundred years. The major market averages plunged more than 40 percent, while the tech-heavy Nasdaq sank by more than 80 percent. The declines in the major averages were the most severe since the 1929–32 bear market, while the decline in the Nasdaq (which didn't exist as an average in 1929) was comparable to the 1929–32 decline in the Dow. When the market began its decline in 2000, stocks were the most important household asset and represented about $16 trillion in wealth. During the market collapse, more than $7 trillion evaporated. This was the greatest loss of wealth not only in U.S. history but surely in any country ever.

Many expected that such unprecedented losses would bring the entire economy down by sending it into a deflationary spiral as consumers, reeling from their stock market losses, drastically curtailed spending. The Federal Reserve was well aware of this risk, which is why it took steps to cut short-term interest rates to practically zero. Such monetary policy meant that banks were practically being given money, and the result was that in some cases, such as zero percent car loans, consumers could take on debt without having to pay any interest. But even zero interest doesn't explain why the economy held firm. If you're feeling desperately poor, as you'd expect all those with huge market losses to feel, you're not likely to take on more debt even if it's interest-free, nor are you likely to do much shopping. Rather, you're going to try to save money for a while.

But this wasn't how consumers reacted in 2000–2002, which is why the economy held firm. One overwhelming positive held back the forces of economic carnage— rising home prices. The ultra-low interest rates made mortgages on new homes exceedingly affordable and sharply increased the demand for homes. Home prices went up across the board.

Wait a minute, you might protest. Didn't we just say that consumers who had lost so much money in stocks weren't likely to go out and spend? Yes, we did. But buying a home is not just a purchase, it's an investment. Over time homes rarely go down in price, and indeed, historically, they have risen in value faster than inflation. In the postwar years home prices have never declined in nominal terms. Thus, purchasing a home is actually better than putting money into a bank, especially when interest rates are so low.

Also important to why the crashing market didn't cause the economy to unravel is the fact that while stocks, in terms of their monetary value, were the most important general class of assets in the economy at the start of 2000, they were more narrowly distributed than were homes. That is, more people owned homes than owned stocks, and thus many homeowners were not directly affected by falling stocks.

Moreover, for homeowners who did suffer big losses in the market, the increase in home values served as a major cushion. First, it buttressed their net worth, and second, it provided a free source of additional capital. That's because lower interest rates engendered a record amount of refinancing, which meant lower mortgage payments. This put more money in homeowners' pockets, which, along with the greater value of their homes, kept them spending. In addition, as homes rose in value, homeowners had easier access to credit. The amount a lending institution is willing to lend is based on the underlying value of the collateral, and in the early 2000s this was going up fast. The bottom line: because increases in home prices can be so easily leveraged, changes in home prices have a much greater impact, dollar for dollar, than do changes in stock prices.

Leveraged Loans

Rising home prices held the economy together in the first two years of the decade. That's the bright side of the equation. The scary side is that falling home prices would cause the economy to unravel. The immense amount of leverage in the economy would operate in reverse gear, with devastating results. Declining home prices would have far worse repercussions than the stock market crash we recently endured. In fact, it is no exaggeration to say that a significant decline in home prices could threaten the foundations of our economic system.

Here's why. Remember, we began by noting that debt in every area of the economy is at record highs. Most of the lenders to whom all this debt is owed are banks. By its nature banking is a very leveraged business, with banks typically lending out many times their equity. For instance, Bank of America's loans typically exceed the bank's net worth by more than six times. This means that if 18 percent of its loans were to go bad, the bank would have no equity left. Unless the Federal Reserve came to its rescue, it would be forced out of business. Nearly all banks in the U.S. are comparably leveraged.

Now, the Fed can rescue one or two banks that are in trouble, as it did in the early 1990s when it kept Citibank and a few other institutions afloat even though by any reasonable accounting standards they were broke. But there's no way it could bail out the entire banking industry if every bank failed at the same time.

The bad news wouldn't stop with banks. If banks were in trouble, it would have a devastating effect on yet another highly leveraged segment of our economy—the corporate sector. In the early 2000s, even with banks in good shape, many leveraged companies, from merchant utilities to some telecoms, were hurting because of high debt loads. Banks that are reluctant to extend loans could cripple these companies in a big way.

What would cause such a large-scale breakdown of the banking industry? A massive surge in bankruptcies. And what could lead to a massive increase in bankruptcies? A huge decline in the value of homes.

If home prices were to fall sharply, consumers would become truly reluctant to spend, in a way that they weren't even during the worst days of the declining stock market in 2000–2002. As we said earlier, when the market declined in 2000–2002, the deleterious effects were offset to a large degree by the increase in home prices. If home prices plunge, it's hard to imagine any comparable offsetting factor coming to the economy's rescue.

A relevant statistic in this regard is the one we cited earlier concerning the huge percentage of loans secured by homes. Declining home prices would force many of those loans, including loans on homes themselves, to be called in. Many homeowners would find that the only way they could satisfy demands to repay loans would be to sell their house. As more houses came on the market, home prices would decline further. General spending would decline as stressed and strapped consumers further tightened their belts.

Banks, homes, spending, the economy, and for good measure almost certainly the stock market, would be caught up in a deadly decline, with losses in each area spurring additional losses in all other areas. Massive consumer bankruptcies would follow, which in turn would put a further squeeze on banks. More loans would be called in, and the vicious circle would become more vicious still. We're not talking recession here—we're talking depression—or rather, Depression.

The Need to Inflate

If you've followed everything we've said so far, you'll realize that there is one key question still to be answered. And that is: what would trigger the large decline in home prices that would lead to widespread catastrophe? After all, in modern times, such a decline is unprecedented.

The one thing most likely to cause home prices to contract, in fact, that almost certainly would do so, would be *policies designed to curb inflation*. Or to say the same thing somewhat differently, it would be policies designed to slow down economic growth—to sharply curb demand for a whole range of goods.

So you see, by way of our discussion of the stranglehold debt has on our economy, we've come around to the issue of economic demand, the counterpoint to the issue of declining supplies of oil that we dealt with last chapter. Remember, demand for goods of any sort is inextricably linked to demand for oil, because oil is involved in every aspect of producing and transporting goods.

Again, go back to the early 2000s and think about what led to the rise in home prices that occurred. It was a steep drop in interest rates. Indeed, thanks to interest rate cuts in 2000–2002, the after-tax cost of borrowing to buy a home was considerably less than the average annual appreciation in home prices. In effect, when you bought a

home you were making money from day one plus securing a place to live.

If rates were raised sharply, though—one of the only
available remedies to bring down inflation caused by
steep rises in oil prices—everything could work in reverse, like a video run backward. In 1980–82, the Federal
Reserve raised interest rates sharply and repeatedly.
While home prices still rose slightly in nominal terms,
they lost value in real terms. The impact of higher rates
on the economy was profound, with unemployment
reaching more than 10 percent. But we survived, and the
stage was set for a long period of economic growth.

In those days, though, homeowners were far less
leveraged than today. Home mortgages represented just
27 percent of the value of homes, compared to more than
40 percent today. In effect, homes today are 50 percent
more leveraged than they were then. No one can say for
sure at what point more leverage ensures that rising rates
would bring on the kind of vicious circle we described
above. What is clear is that policymakers don't have
the same options they used to have to combat inflation.
They can't blithely restrict money and raise rates. They
wouldn't and shouldn't dare.

In fact, that's not even the whole of it. It's not just that
policymakers will have to tolerate high inflation—they
actually will have to take steps that will foster it. That's
because if, for any reason, the economy started to
weaken, policymakers would have to react vigorously or
risk triggering a downward spiral. And high energy
prices, as we pointed out earlier, are in and of themselves
an economic depressant. So when oil prices climb,
policymakers will be forced to provide the easier money

that might offset their effects on the economy—to cut taxes or loosen monetary policy.

Once inflation starts to rise, policymakers will be in the position of a doctor trying to treat a person with an inflammatory condition, say, arthritis, that can be treated only with strong doses of anti-inflammatory drugs. The problem is that the drugs cause bleeding ulcers. The doctor in charge has to decide which is worse, the inflammation or the ulcers. If the inflammation is merely uncomfortable, but the ulcers will kill, there's really not much choice.

In the world of medicine, there is always the hope that research will come up with new medicines that will cure without unpleasant or deadly side effects. In the world of economics, however, such miraculous solutions aren't available. We're stuck with all the existing relationships that will continue to rule. And this means we'll be powerless to fight inflation. When energy prices take off, we won't be able to bring them down—at least not until we develop alternative energies, which would be the economic equivalent of a new, side-effect-free miracle drug.

There's one final factor that will weigh in on the side of inflation, and that's the self-interest of our political leaders. Remember the motto of Bill Clinton's first presidential campaign: "It's the economy, stupid"? It could have been refined even further, to "It's the recession, stupid." It is well-established political dogma that nothing puts voters more in the mood to kick out incumbents than a nasty recession. Given the choice between policies that will rein in inflation but cause a major economic contraction and policies that will encourage inflation but also

spur economic growth, politicians with a healthy dose of
survival instincts aren't going to find it much of a contest.

Key Points:

◆ Debt in the U.S. is at record levels—2.8 times the
  gross domestic product, a level not seen since the
  Great Depression. Consumer debt has been the fastest-
  growing segment.

◆ When an economy is loaded with debt, strong economic growth becomes essential, even at the price of
  high inflation.

◆ Most of the huge overhang of consumer debt—about
  70 percent—is in the form of home mortgages, and a
  huge proportion of all consumer debt is secured by
  the home. The home is central to consumer finances.

◆ Strong home prices kept the economy going after the
  crash in tech. Falling home prices would cause the
  economy to unravel.

◆ As inflation rises, the Fed will find its hands tied because efforts to curb inflation by raising interest
  rates would hit at home prices. In our highly leveraged economy, this would set off a devastating
  downward spiral, pulling down consumers, banks,
  and corporations.

CHAPTER 6

Moving Beyond Oil, Part 1

I t could be, and for all we know is, the title of a pop song—"Where do we go from here?" Our analysis has taken us right to the edge of an impasse. We'll need more oil, and there will be less of it around. How will this dilemma be resolved?

Actually, given that this is primarily an investment book, we could duck this issue altogether. We could simply make the case for an upcoming oil crisis, note the most likely economic effect—inflation—and proceed straight to part II, where we present the best investments and hedges for turbulent and largely inflationary times.

That wouldn't satisfy us, though. We're willing to spend the time and make the effort to look at possible solutions to the oil squeeze. Not only do we have opinions about what will happen, we also have ideas about what should happen—what policies we should try to implement. And this is no time for anyone to be shy about expressing views on this topic. How well this country deals with the growing inability of oil to meet our energy needs will affect not just us but also our children and grandchildren and great-grandchildren. It is slated to emerge as one of the dominant issues of our times. And it has a wide range of serious investment implications. We think it's important that everyone understands as clearly as possible both what is at stake and what the various choices are.

Logically, if we're right that we will be unable to find enough new oil, there are really just two approaches to take. First, we can try to conserve our remaining oil by using it far more efficiently than we have up to now, stretching out the amount of time we have before we really feel the crunch. Second, we can substitute other fuels for oil—whether fuels that we already rely on in part or new fuels that we attempt to develop.

In this chapter and the next we look at these possibilities in some detail to see which make the most sense. We examine a range of alternative fuels—both renewable and nonrenewable—their pros and cons, availability and degree of readiness, and look at conservation.

On one level, it's a quick survey course on oil alternatives. But as we indicated, we have strong views on the subject, and in the following pages we're not just analysts, we're advocates. In particular, next chapter we offer the basics of what we think is a novel, feasible, and, while expensive, essential proposal for moving beyond oil, one that involves a combination of "old" and "new" energies.

In this chapter we look at the extent to which it may be possible or smart to increase our use of "old" energies to help make up for the oil gap. Next chapter we turn to the energy sources people generally think of when they think of alternative energy, that is, energies that are non-polluting and infinitely renewable.

A final note: we really haven't lost sight of your investment health. While these chapters look at various energy sources largely from a science and policy perspective, they also are laying the groundwork for recommendations in this area, presented in chapter 13.

Natural Gas

Oil production may be on the verge of peaking, but oil isn't the only fossil fuel. Coal and natural gas are still around, and we still use a lot of each one. And while they aren't what you usually think of when you think of alternative fuels, they actually will both have to play a role in the coming transition away from oil and toward renewable energies.

All fossil fuels are predominantly a combination of carbon and hydrogen. When we burn a fossil fuel to release energy, we are burning the hydrogen, leaving carbon as the by-product. Coal has the greatest proportion of carbon, which is why it is the dirtiest, most polluting of the fossil fuels. Natural gas has the lowest proportion of carbon, which is why it is the cleanest-burning fossil fuel.

Why can't we just substitute natural gas, the most environmentally friendly of fossil fuels, for disappearing oil? There are several reasons. First, like oil, natural gas is also in short supply, at least in the U.S. There are compelling signs that our endowment of natural gas is peaking, even as demand is rising. Because of its low carbon content, natural gas has become the fuel of choice for new electric generating plants. Over the next decade, demand for natural gas in the U.S. is expected to increase by over 30 percent, or by about 3 percent a year. There is no evidence that we will be able to satisfy this demand.

Figure 6a, "Natural Gas Prices," tells the story. The years from 1990 to 2000 bear a strong resemblance to the oil chart in chapter 1, except that prices fluctuated around $2 (per 10,000 cubic feet) rather than $20 a barrel. Every time prices climbed above $2 they would quickly move back below that level. The highest prices came during the very cold winter of 1996, when storage levels fell to what was then a record low.

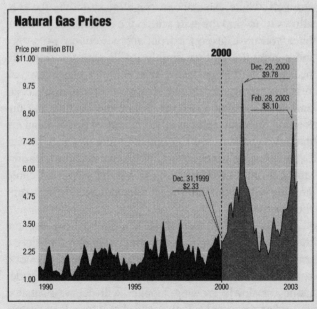

Natural Gas Prices

Price per million BTU

2000

Dec. 29, 2000
$9.78

Feb. 28, 2003
$8.10

Dec. 31, 1999
$2.33

$11.00
9.75
8.50
7.25
6.00
4.75
3.50
2.25
1.00

1990   1995   2000   2003

Figure 6a

But in the early 2000s, despite the fact that economic growth in the U.S. slowed down, natural gas prices averaged nearly $4, roughly 100 percent higher. The $2 average price of the 1990s had become the low price in the 2000s.

So as the balance between supply and demand for oil was shifting in 2000, so was the balance between supply and demand for natural gas. It is clear that we will have to rely on heroic efforts to come even close to meeting our need for natural gas. Even after prices surged in 2000, generating record drilling, natural gas production increased just a bit more than 2 percent in 2001, not enough to meet one year's projected average growth in demand.

And unlike oil, natural gas can't be easily transported from elsewhere. An elaborate pipeline system in the U.S. allows us to ship the stuff within the continental U.S. But once you try to export it from another country, it's a lot more complicated than simply loading it onto ships. To transport natural gas other than through pipelines, you first have to liquefy it, which involves considerable costs and risks. Natural gas ports, where the loading and unloading are done and gas is stored, are expensive and are natural hazards as well as potential targets for terrorists. On a best-case basis, according to a recent Smith Barney study, an all-out effort to build natural gas ports could provide only about half the incremental need for gas over the next decade.

In sum, like oil, natural gas is a diminishing resource that offers no way to increase supply enough to satisfy a growing demand for energy. Natural gas clearly is not a solution to the looming energy crisis. But the rising demand for slowly growing supplies will put well-situated natural gas companies in the catbird seat for a long time to come. As we'll discuss in chapter 11, they will be among the best investments you can make in the energy patch.

Coal

For coal, the issues are very different. Supplies of coal are abundant, both in the U.S. and worldwide. A major portion of our electricity still comes from coal-fired plants. The problem, as we said, is that coal is dirty, and technology hasn't come up with very good or efficient ways to clean up the stuff, at least not at a price that makes sense. When we burn coal, it emits high levels of pollutants that do such nasty things as hasten global warming and foul the environment, not to mention threatening the health of those who mine it.

But if the energy crisis we face really is so critical to our survival, isn't some environmental damage and even slightly higher death rates a reasonable trade-off? There are good reasons to answer no.

Many consider coal to be the most significant environmental threat the planet faces, because coal emits more greenhouse gases than other fuels. The *Economist,* known for its hardheaded assessments of world issues, featured coal on the cover of its July 6, 2002, issue with the caption "Environmental Enemy No. 1." The magazine argued that it is essential that we cut back on coal use as much as possible, noting that "warming may trigger irreversible changes that transform the earth into a largely uninhabitable environment."

True, no one can prove that global warming exists, much less that it will cause severe damage worldwide. There are temperature trends and lots of scientific analysis, and there is an overwhelming consensus among scientists that the planet is warming. But before Einstein there was an overwhelming consensus that the speed of light was variable in a vacuum, and before Copernicus there was a consensus that the earth was the center of the universe. Should the current consensus be enough to make us play down the one energy source that we have in such abundance?

In answering this question, there is one number that stands out and that suggests the world's environment is indeed becoming much more turbulent and that we should pay attention. That is the stunning increase in weather-induced property damage that has occurred in recent years. During the past decade, property damage exceeded all the weather-related damage sustained worldwide in the four preceding decades combined. That's an astounding increase. And while we know that there is no absolute proof that increased pollution and global warming are the culprit, it would be taking a massive risk to ignore these trends.

If we ascribe just a small portion of these damages to the burning of coal, the actual cost of coal begins to skyrocket. If property damage were to take another major leap upward, it could deal a body blow to our economy from which it might not recover.

Another quick investment preview: rising property damage has a direct impact on one group of investments, the property and casualty insurers, whom we discuss in chapter 15. It hurts some and gives a tremendous boost to others. One of our prime investment picks comes from this group.

Coal presents one other major though lesser-known problem: it can contribute to nuclear proliferation. One good-sized coal-burning electric plant releases enough radioactive material in a year to build two atomic bombs. This very serious problem was pointed up in an article in *Foreign Affairs* in January/February 2000. Authors Richard Rhodes and Denis Beller quote physicist Alex Gabbard, who noted "the collection and processing of coal ash for recovery of minerals . . . can proceed without attracting outside attention, concern, or intervention. Any country with coal-fired plants could collect combustion by-products and amass sufficient nuclear weapons materials to build up a very powerful arsenal." The last thing the world needs is more nuclear threats.

In sum, while coal will continue to be needed for a while longer, there are too many strikes against it for us to be willing to accept a situation where we are forced to increase our use of coal. Rather, over the long term we need to reduce our dependence on coal.

Nuclear Fission

Nuclear reactors could, possibly, solve all our energy problems—if nuclear energy didn't have so many problems of its own. There are two types of nuclear reactions: fission and fusion. In fission, nuclei are forced apart, in the process releasing tremendous energy. In fusion, energy is released when nuclei are forced together. We can get lots of energy through fission, but we don't know what to do with the radioactive wastes that are its by-product. Fusion, which we'll look at next chapter, doesn't have those safety issues, as its only by-product is water—but we are nowhere near figuring out how to carry out fusion so as to get a net gain in energy.

Nuclear plants operate in countries around the world, and in some, such as France, they contribute the bulk of electric power. In the U.S., nuclear reactors relying on fission currently supply about 20 percent of our electricity and about 10 percent of our overall power. Worldwide, nuclear fission accounts for about 7 percent of all energy usage.

Some analysts have argued that nuclear fission offers the best hope for freeing ourselves from our dependence on fossil fuels and urge a crash program to build more reactors. They point to the relatively low cost of electricity from nuclear reactors and to the improving safety record. But they ignore some serious realities that almost certainly will, and in our view should, limit the role of nuclear fission to primarily a circumscribed and short-term fix.

First, while it's true that once nuclear power plants are in place the electricity they generate is relatively low-cost, this ignores the massive expense of design and construction. In fact, there is a reasonable argument that once you factor in these costs, nuclear energy is an immoderately expensive form of power.

Second, the safety issues are both immense and nowhere near to being solved, and in fact, there is good reason to believe they may not be solvable. To understand just how thorny they are, a brief physics lesson is useful. Virtually everything we see and touch—wood, water, people—is held together by chemical bonds known as the electromagnetic force. Obviously these bonds are very strong or we would all be flying apart. However, they pale in comparison to the bonds that hold the nucleus of an atom together. This force, known as the strong force, is many times stronger than the electromagnetic force.

The reason that nuclear reactions generate so much energy is that the energy released is proportional to the force involved, which by a huge margin is the strongest force in nature. The by-products of this reaction are highly radioactive rods and other material, which remain dangerous to human beings for thousands of years as they very slowly disintegrate. These radioactive by-products need to be contained, prevented from leaking into the environment, where they would play havoc with human health. Currently they sit in what are basically makeshift containers on the site of the nuclear plants that generate them. But because these radioactive contaminants are products of the strong force, they will eventually overwhelm any container, whose bonds are so much less strong.

To cope with all this radioactive waste, all that the best minds have been able to come up with is to bury it deep in the ground. But where? As you probably know, Yucca Mountain in Nevada has been granted that dubious honor and is slated to receive some 70,000 tons of existing nuclear waste. Even after twenty years of study, though, this remains a controversial choice. R. C. Ewing, a nuclear engineer from the University of Michigan, wrote in the April 26, 2002, issue of *Science*: "At Yucca Mountain, the passive properties of the repository site do not provide a long-term barrier to radionuclide release." And he noted other unresolved issues as well—including the possibility of volcanic activity. In fact, it's worth noting that over a long period of time it's impossible to rule out earthquakes anywhere on earth. The most powerful earthquake ever recorded in the U.S. took place more than a century ago in the Midwest, and to this day no one has explained how this happened, given that there is no fault line to account for this extreme event. The possibility of an errant volcano certainly makes the thought of thousands of tons of underground nuclear waste an unsettling prospect.

And, of course, the current debate over Yucca Mountain would intensify many times over if we ever stepped up the use of nuclear reactors, generating ever-expanding piles of radioactive waste. Closing our eyes to this problem would be as woefully shortsighted as our blindness over the past many years to the need to develop energy alternatives.

Radioactive waste is just one of several safety issues inherent in nuclear reactors. Another is the proliferation of nuclear weapons. Both Pakistan and India developed their nuclear weapons programs in conjunction with nuclear reactors, and North Korea is following suit. Another risk is the potential for terrorist attacks on nuclear facilities. And a final risk is of accidents such as the ones at Three Mile Island and Chernobyl. These have been rare up to now. But if nuclear energy became more widespread, this would change. An article in the May 19, 2000, issue of *Science* noted: "In a world with about 4000 reactors . . . the expectation would then be for a TMI-scale nuclear accident every several years."

For all these reasons, nuclear fission isn't a long-term solution to the world's need for energy. Short term, though, it will remain an important stopgap. The increased use of nuclear energy was a big part of why oil prices declined in the 1980s and 1990s. The nuclear energy produced today worldwide far exceeds the world's excess oil capacity and, at least in the U.S., its excess natural gas capacity as well.

It's a sad testimony to our energy capabilities that

right now the only significant alternative to fossil fuels is nuclear fission, a technology that has no future and that in many ways is far less palatable than fossil fuels. But that's how it is. One implication for investors is that nuclear utilities are good hedges against the inevitably rising costs of fossil fuels. We'll have more on this in chapter 11.

Hydroelectric

One other energy source that makes a modest—very modest—contribution worldwide is hydroelectric. Hydroelectric power converts the energy from falling water into electricity. To use hydroelectric power, you need a source of water, fairly sharp differences in elevation, and dams so that you can store water at higher elevations. Thus, hydroelectric power clearly isn't something feasible for most parts of the world, and it's not going to be a growing source of energy. In addition, when you build dams, you create a variety of environmental problems, including the release of greenhouse gases. Hydroelectric power accounts for only 6.9 percent of worldwide energy usage, up marginally from the 6.3 percent figure a decade ago. In the U.S., hydroelectric power has fallen off sharply and today is barely on a par with wood.

The Trouble with Conservation

A few years ago, Vice President Cheney enraged liberals and environmentalists when he rather patronizingly commented that while conserving energy might be a sign of personal virtue, it wouldn't do much to solve the nation's energy problems. A storm of indignant editorials and letters to editors denounced his comments, and facts and figures were offered showing just how many barrels of oil could be saved through higher mileage requirements for cars and other fuel-efficiency measures. The numbers were impressive.

Still, while the purpose behind Cheney's bad-mouthing of conservation was to promote oil exploration and drilling as the answer to our energy needs, in a way he was right—not about the drilling but about gung-ho efforts to force the use of less oil through government mandates or through subsidies to households that conserve. The problem is that to the extent that these efforts succeed, the effect is to keep oil prices lower than they otherwise would be—after all, you're cutting demand. And we believe that our future is viable only if we develop renewable alternative energies, the sooner the better—which will happen only to the extent that oil prices rise. So while alternative energies and conservation seem linked, and although proponents of each— generally the same people—share the same ultimate goal, which is to cut our use of fossil fuels, there actually is an inherent tension between the two approaches. Over the long haul, conservation can only put off the day of reckoning, while the development of true renewable energies will solve the problem pretty much forever.

It's also possible that conservation might simply cause us to spin our wheels. That is, to the extent that gains in fuel efficiency keep the lid on oil prices, consumers are likely be less frugal about oil for other uses, such as home heating—the oil savings from conservation will be at least partially squandered elsewhere. Still, conservation is likely to have the overall effect of keeping oil prices lower than they otherwise would be. And while this might sound like a good thing, it isn't in our best interest.

Why can't we do both at once? That would seem to be the best of all worlds, but the fear is that it just wouldn't work out that way. It's like asking why we didn't develop alternatives when oil was at $4 a barrel. In a free market system, price is always going to be the chief determining factor. Until oil prices rise sufficiently, we're going to keep putting off facing reality.

And that could be exceedingly dangerous. One reason for a sense of urgency about zeroing in on alternative energies as rapidly as possible, rather than dragging out the process by efforts at conservation, is that we can never know when we will experience a sudden oil shock that will abruptly choke off supplies. Unlike the 1970s, when worldwide oil supplies were more abundant, we're too close to the edge now to be sure we can survive such a blow. And with the world as fragile as it is, with most oil concentrated in politically hostile or unstable countries, it obviously would be naive to say that such a shock won't happen.

Whatever your view of the primary motives behind our unprecedented initiation of the war in Iraq, this reality—the importance of securing world oil supplies—surely wasn't entirely absent from the minds of our leaders. In later chapters we talk about how our vulnerability to oil shortages makes it essential that we continue to build up our military might. But trying to police the oil-producing world through our soldiers and weapons will never be a hundred percent foolproof—oil shocks can still occur—and efforts to do so obviously are fraught with a great many risks and problems. The only long-term answer is to develop acceptable and viable alternative forms of energy, and the sooner we do it, the better.

So yes, it does seem disgraceful that the average mileage achieved by cars has been dropping significantly, one of the statistics frequently pointed out by pro-conservationists. But when oil prices rise enough, SUVs will fall out of favor without a nudge from the government. Seeking to force conservation artificially is merely likely to put off that day.

Key Points:

- Lagging oil supplies will force us to turn to other sources of energy.
- Natural gas, the cleanest-burning fossil fuel, won't be a ready substitute because it, too, is in increasingly short supply in this country. Transporting it from elsewhere poses major risks and problems.
- Coal is abundant but dirty. Nuclear power may temporarily help fill the gap, but its risks make it unacceptable as a long-term solution.
- Efforts to mandate conservation will only put off the day of reckoning when we need to move beyond oil, and they may lull us into complacency.

CHAPTER 7

Moving Beyond Oil, Part 2

Recently there was a fascinating article in the *New York Times* about a restaurant owner who reconfigured his car—a Ford Excursion with a turbo-diesel engine—so that it would run on the waste vegetable oil his restaurant used for cooking french fries. In essence, the fuel was free, because otherwise the oil just would have been thrown away at the end of the evening. The car apparently ran as smoothly as ever, and no harmful pollutants were released into the air. The only drawback, if you can call it that, was that the car's exhaust smelled faintly of french fries.

When you're writing a book about energy, you tend to notice stories like that. And it's entirely possible that other restaurant owners may take to using vegetable oil, or even olive oil, in their cars' gas tanks. However, in general when you're talking about alternatives to petroleum, cooking oils rarely come up. Clearly they're niche fuels, and we'll say nothing more about them.

Getting back to more conventional fuels, the energies discussed last chapter all have inherent problems. They're dirty—coal; they're dangerous—nuclear fission and to some extent coal; they're running out—natural gas; they're of limited use—hydroelectric. A sorry bunch indeed.

So now we turn to the energies that seem to better fit the common notion of what we mean when we talk about alternative energies. These are the energies that come from limitless natural resources, that are renewable and seemingly nonpolluting. Wind, sun, hydrogen from water—from these, it is hoped, will come the energies of the future. If we could harness the wind effectively enough, we wouldn't need the Middle East. If we could capture sunlight efficiently enough, we wouldn't have to worry about global warming. Nor, come to think of it, would we have to worry about economic growth or inflation. A lot of problems would be solved.

It's not that these fuels are all brand-new, not by a long shot. Some have been in use in one form or another for a long time—think of windmills in the Netherlands—and are helping us meet our energy needs today. But they all need a huge investment of money, either to create the necessary infrastructure for more widespread use or to conquer big technological hurdles. In other words, these alternative energies aren't yet ready to take center stage. But make no mistake about it—if we don't commit ourselves to a massive effort to develop at least some of these fuels, we're going to be in a lot of trouble. Below we assess how far along we are with these energies, what it will take to get them up to speed, and which ones we should focus on most seriously to get us through the coming decades.

The transition won't be simple. Right now clean renewable energies account for only a small fraction of one percent of energy usage in the U.S., about the same percentage as ten years ago. In the next few decades, the level will have to increase exponentially. That's a very big deal with major implications for the economy and investors.

Solar

When it comes to shifting to clean, infinitely renewable energies, probably the most extravagant hopes attach to the possibility of using sunlight to power the earth. It has been estimated that enough sunlight hits the earth each day to meet the world's energy needs for twenty years. The trick is to capture it and store it for later use, and to do so sufficiently cheaply in terms of the money, energy, and natural resources expended. We've made progress in tackling these problems, but a lot more research is needed to make solar energy feasible or likely on a large scale.

The technology to harness sunlight and convert it into electricity is built around solar cells, also known as photovoltaic, or PV, cells. The idea is that sunlight is directed onto light-absorbing material in such a way as to "excite" that substance's electrons and result in electrical power. Solar cells were first used in the late nineteenth century as light meters in photography. In the 1950s scientists at Bell Labs took the technology a lot further. Using silicon, they produced cells that could convert 4 percent of the energy in sunlight into electricity. Within fairly short order solar cells were being used in the space program.

Today solar energy is also used in a wide variety of other applications, from solar-powered calculators and watches to emergency radios, lighting, and the pumping of water to solar panels on the roofs of buildings. On sunny days these panels can capture the light that falls on them and convert it into electricity that can turn on electric lights, run dishwashers and air conditioners and other appliances, and so on. Such solar panels are often used in areas of the country that are far from central electrical generating facilities. They also can be connected to a utility-serviced electric grid. In that case, solar energy becomes the energy source when the sun shines; the grid takes over at other times. Excess energy produced by the solar panels can be transmitted to the utility for dispersal to other customers—further reducing fossil fuel usage and helping to defray costs for the solar-panel-owning homeowners.

Improving the cost-effectiveness of photovoltaic cells is essential if solar energy is to play a major role in the future. Two things are necessary: bringing down the cost of making solar cells and, more important, improving their efficiency so that they convert a significantly larger proportion of light to electricity. If and when we make enough strides in these directions, we'll be close to being home free.

So far three basic types of solar cells have been developed. Without getting too bogged down in detail, the important thing to know is that there are trade-offs between how expensive they are to manufacture, how wasteful their manufacturing process is of natural resources, and how efficiently they convert sunlight into electricity. No one method hits the jackpot on all three counts. Given the recent rate of progress in efficiencies and costs, which has averaged about 5 percent a year, it will take at least thirty more years for solar cells to be competitive with natural gas and coal, and possibly much longer. That's assuming that the federal government doesn't step in with massive funding to push the technology along faster. But even if it does, there is no guarantee of results.

Apart from the fact that they cost a lot to make and don't convert sunlight to energy efficiently enough, there are a couple of other problems with today's photovoltaic cells. First, the manufacturing process results in toxic wastes. Second, making them requires the extensive use of certain natural resources, such as iron. Rhodes and Beller, in the *Foreign Affairs* article we cited last chapter, argue that a global solar energy system would take a century to build and would consume a major chunk of world iron production. It seems pretty clear that given the existing technology, solar energy is nowhere near being ready to meet the world's need for energy.

One expert in this area, John Turner, writing in the July 30, 1999, issue of *Science,* takes a different position, arguing that it is theoretically feasible for photovoltaic cells, under existing technology, to meet all the energy needs of the U.S., both for electricity and for transportation. Unfortunately, however, his advocacy of solar energy only serves to indicate how far away we really are from a solar world. For one thing, under his calculations, to site the freestanding solar cells needed for solar to take center stage would entail setting aside an area the size of Nevada. Turner notes that while this is a large area, it is less than one-fourth the area the country has devoted to roads and streets.

But let's say that we were willing to set aside a Nevada-sized clearing somewhere in this country, or even to level Las Vegas and its environs and replace them with solar panels, something some people might consider a reasonable trade-off (just kidding). Turner's argument is still disingenuous, because it is based on an assumption that we have achieved something we haven't—the ability to use the sun to crack water efficiently so as to obtain pure hydrogen as a fuel.

Hydrogen

It's the Holy Grail of energy research. You take water, any water, from anywhere. You take sunlight, the sunlight that streams down to the earth every day. You use the sunlight to split water molecules, freeing up hydrogen. You then burn the hydrogen, meaning that you recombine it with oxygen, to generate energy to run factories and power homes and cars. If you can, in the lingo of alternative energy, crack water with sunlight, you have a source of cheap, limitless energy that adds zero pollutants to the earth. What a deal!

Hydrogen is a really frustrating element. It's the most abundant element in the universe, and when you get it on its own, it's an ideal fuel. It can be burned directly in internal combustion engines, where it is far more efficient and less polluting than gasoline, or it can be fed into fuel cells, where it is chemically combined with oxygen to produce energy. Either way, it produces scads of power and the only by-product is water. The problem, though, is that hydrogen isn't found on its own. It has to be isolated, separated from other elements. And doing that efficiently and affordably is the rub.

As we noted last chapter, in fossil fuels hydrogen is combined with carbon. When we burn fossil fuels, we haven't attempted to isolate hydrogen first—we're burning the whole thing, which results in dirty carbon emissions and in some cases other toxic materials. There are experimental efforts to separate the hydrogen from the carbon in coal before using the hydrogen as a fuel. But so far it's very expensive, and you're also left with the problem of safely disposing of the carbon dioxide that results.

Hydrogen that comes from water doesn't have that drawback, but again, the problem is isolating it affordably, splitting it off from the oxygen with which it is bound. To do that, you have to apply energy, and that energy has to come from someplace else—whether from the sun or fossil fuels or some other source.

And this brings us to the symbiotic connection between the sun and hydrogen. Hydrogen is the crucial missing link in the search to obtain maximum benefit from the sun's potential as a fuel. The reason is that without an intermediary medium like hydrogen, the energy from the sun can't, in effect, be stored for future use, or transported from one place to another. We can use solar panels, as described above, to generate electricity for a lot of purposes, and they can certainly play a role in weaning us from fossil fuels. But without a storage medium like hydrogen, their role is limited. We can't use the sun directly to run cars, for instance, or easily use energy from photovoltaic cells to do work at too great a distance away. If the sun could be used to crack water, however, the hydrogen thereby obtained would be available to do all those things.

Research in this area is focused on experimenting with various combinations of metals in order to come up with one that can efficiently absorb the sun's energy and apply it to separating H_2 from H_2O—cracking water. There is nothing theoretically impossible about this. But so far, the materials that are able to split water don't absorb sunlight effectively enough to be useful or economical, converting only 1 to 2 percent of the energy from sunlight into stored chemical fuel. The materials that absorb sunlight more efficiently aren't constituted in such a way as to be able to split water. This is one roadblock to achieving a true hydrogen economy.

There is a second problem: we need to make giant strides in our ability to store and transport hydrogen safely and efficiently. Hydrogen is a relatively unstable element, and you can't just put it into, say, a canister and carry it around. Research is being done on creating lightweight but safe composite materials more suitable for hydrogen storage than anything we have now. And that's still not all. If and when such materials are developed, the next issue would be building the large amount of infrastructure needed to have hydrogen available for widespread use.

In sum, it seems to us that a real solar-based hydrogen economy is a long way off, many decades away at best. We're not dismissing it as an ultimate possibility and think it's an area that merits significant funding for research, but we certainly don't think there is any possibility that it will be up and running in time to help us deal with the oil shortages and rising oil prices that lie in our more immediate future. We'll need to find another answer to get us through the coming period.

Hydrogen Car Hoopla

If over the past few years you've paid any attention to developments on the renewable energy front, you may dimly recall that in 2003 there was some hoopla over hydrogen cars. In fact, the front page of the March 5 business section of the *New York Times* had the headline, "Hydrogen Vans and Pumps Head for Washington." And President Bush in his 2004 budget proposal included a $1.7 billion subsidy for research into cars powered by hydrogen fuel cells. Maybe we're being unduly pessimistic, and a hydrogen economy is closer at hand than we think?

Not exactly. First, the touted General Motors hydrogen cars cost around $5 million each. GM's vice president for research, Larry Burns, said he was optimistic that the company could produce an affordable car running on a hydrogen fuel cell by 2010, though he admitted it would be a challenge. And maybe it will happen—but it would be an amazing feat. To bring down the price from $5 million to the tens of thousands in such a short period of time would require economies of scale that would dwarf anything ever accomplished even by the electronics industry. In other words, to meet this goal, the automotive industry would have to do something no industry ever has done before.

Second, even if General Motors could produce such a car, it's important to realize that when you hear talk today of hydrogen-powered cars, it is assumed that the hydrogen is obtained from natural gas, the same natural gas that we're arguing will be in increasingly short supply. In addition, fuel cells require as catalysts other resources that also are in limited supply, such as platinum. To the extent that fuel cells become common in cars, platinum prices will go up, making these cars more expensive. And other issues remain relevant—such as the need to develop a better storage medium for hydrogen and to create a nationwide infrastructure so that drivers could count on being able to fuel up anywhere, anytime.

On the plus side, hydrogen fuel cells generate power far more efficiently than the internal combustion engine, so overall they are likely to be energy-saving. Moreover, they are nonpolluting. But they are not around the corner, at least not in any great numbers, and they won't single-handedly solve, or even begin to make a real dent in, our energy problems over the next decade or two.

Fusion

The sun is an enormous nuclear fusion reactor, generating energy by continually forcing the nuclei of hydrogen atoms to combine. Thus, indirectly, nuclear fusion is the source of all the energy we have on earth. But though we've been trying to duplicate the processes that go on in the sun, we haven't had any luck so far. Or to be more precise, we know how to carry out fusion—we just don't know how to do it so as to create a net gain in energy. It takes more energy to force the nuclei of different atoms to fuse than you end up getting from that fusion. Not much of a bargain. It's as if an alchemist swore he had a secret recipe for making gold, but it turned out that to make two ounces of gold, he had to put in three ounces. And it's a shame, because fusion is clean.

How far away are we from actually creating energy through fusion? A long way. For nearly twenty years a consortium of nations—not always the same ones—has been trying to raise funds and support for what is considered to be the best chance of achieving fusion in an experimental situation. The project is called the International Thermonuclear Experimental Reactor (ITER). As of this writing, the U.S., after having dropped out of the consortium for several years, has agreed to rejoin. What originally was a $10 billion project has been scaled back to $5 billion. Dennis Normile, writing in the February 28, 2003, issue of *Science,* sums up the prospects: "The heart of ITER is a mammoth vacuum vessel surrounded by several types of superconducting coil . . . that will magnetically confine a hydrogen plasma in the shape of a doughnut. . . . Such experiments are considered the next step toward learning how to exploit fusion as a source of energy. In addition to the main reactor, the complex will include a dozen or so buildings and structures spread over 30 hectares. Construction is expected to take 10 years." In other words, once they begin construction it is slated to be a decade— and we've never known a construction project that didn't take longer than expected—before they even will be in a position to evaluate those next steps. In sum, fusion is not something we're going to see anytime soon. It certainly is not going to get us through this decade's critical transition away from oil.

Wind

The great advantage to wind energy is that the technology for it is a lot simpler and a lot further along than for solar energy. It involves building wind turbines, high towers with rotating blades, in areas that typically are exposed to a fair amount of wind. The wind turns the blades, and the energy is ultimately transmitted to a generating station that can disperse it to end users. A certain density of turbines is required for wind energy to be effective, and thus the turbines are sited in groups on wind farms.

One objection sometimes raised to wind turbines is that they could lead to the wholesale decapitation of birds flying into the rotating blades. It may be possible in most cases, though, to site the turbines so that they are out of the way of avian migration paths. On the plus side, the land on which the turbines are located can simultaneously be used for other purposes, such as certain kinds of farming. Turbines also can be sited offshore in the oceans.

Wind energy is catching on in Europe. As of the year 2000, Germany had 6113 megawatts of installed turbines (roughly 2 percent of its total electrical output), and Denmark had 2300. This is a minuscule portion of their overall energy capabilities, but it is still significantly more than in the U.S., which has just 2254 megawatts of installed wind turbines. Outside the U.S., the use of wind as an energy source has been growing by more than 20 percent a year. Inside the U.S., growth has been only in the mid single digits—and that is starting from a far smaller, not to say nominal, base.

So the technology for wind is in place—no question about it. The only real issue involves the costs, and here, too, wind energy appears to be a winner. There is no question that the costs of wind energy have come down dramatically. In an August 24, 2001, article in *Science,* Mark Jacobson and Gilbert Masters, both members of

Stanford University's Department of Civil and Environmental Engineering, assert that wind energy now costs less than coal. In Denmark, they report, energy from wind turbines, whether large or small, costs 4 cents per kilowatt-hour. "These numbers suggest that the total costs of wind energy are less than those of coal energy . . . every 36,000 to 40,000 turbines could displace 10 percent of U.S. coal at a cost of $61 to $80 billion."

There also are good arguments that wind energy is cheaper than natural gas as well. In their book *Wind Energy in the 21st Century,* published in 2001, authors Robert Y. Redlinger, Per Dannemand Andersen, and Poul Erik Morthorst, referring to a dispute in Minnesota in the late 1990s between a utility and a public interest group, lend support to the view that wind is 7 percent cheaper than gas even apart from any production tax credits that might be offered. It's worth noting that when this case was brought, natural gas prices were less than half of what they rose to in 2000–2002. Moreover, in the past few years wind technology has improved. All this suggests that today wind has a considerable cost advantage over natural gas.

Others concur. In its *State of the World 2003* report, the Worldwatch Institute noted: "During the past 15–20 years, wind energy technology has evolved to the point where it competes with most conventional forms of power generation. In many instances, wind is now the cheapest option on a per-kilowatt basis."

Some scientists, however, have argued that these rosy assessments don't take into account the hidden costs of using wind energy—in particular those related to wind's intermittency (since you can't count on the wind always blowing, you have to have a backup energy source, which costs money) and transmission. (You have to get wind energy from the place where the wind is blowing to electric grids able to reach all parts of the country. Since North Dakota happens to be the windiest state, this clearly can't be ignored.)

Rebutting these arguments, Jacobs and Masters have argued that such costs actually amount to less than 2 percent of the price of wind energy and that they can be reduced even more. Moreover, even those who have questioned wind's cost-competitiveness agree that if instead of looking just at coal's market price you factor in the very real costs of coal's environmental damage, wind is clearly competitive.

Wind and Coal: Our Proposal

Wind theoretically can supply the lion's share of our electricity—but it can't run our cars and trains and planes, so it can't directly solve all the problems of diminishing oil and natural gas. Given this reality, and the fact that neither fusion nor a hydrogen economy based on advanced solar technology is anywhere in the immediate offing, how can we best get through the next decade and beyond? Is there any hope?

We think there is. You may have wondered why in this chapter devoted to "new" energy alternatives, we dared in the above subhead to mention coal, a dirty old energy, in conjunction with wind. And we admit that it is a bit incongruous, reminiscent of those questions on intelligence tests given to four-year-olds where they have to spot what item doesn't belong (pizza, hamburger, baseball bat, french fries). But it wasn't a misprint.

Here is how we think it can play out, our rough-and-ready proposal for finessing the difficult transition from fossil fuels to, we hope, an eventual but necessarily distant future based entirely on clean renewable fuels. The most feasible answer to shrinking oil would be to dramatically increase our use of wind energy, in conjunction with using our abundant coal resources at existing rates but in innovative ways. This approach would enable us to reduce our dependence on oil and natural gas and move us closer to relying on renewable energies. And while its inclusion of coal as a key component might seem retrogressive, the key point is that it wouldn't require an increase in the amount of coal mined, just a redirection of it.

Our suggestion is to begin immediately to take steps so that wind energy can generate a growing proportion of our electricity, taking this job over from coal and natural gas. Right now, coal generates about 50 percent of the electricity used in the U.S., and natural gas accounts for a major—and growing—chunk of the remainder. Wind could take over the lion's share.

This would free up a whole lot of natural gas and coal, which then could be devoted to jobs now done by oil—like fueling our cars. We noted earlier that one problem with today's hydrogen fuel cells is that they rely on natural gas, which is in short supply. But if natural gas no longer had to supply us with electricity, it would be more readily available for use in fuel cells at whatever point hydrogen cars begin to make inroads in the automobile market.

As for coal, the technology exists today, and is being improved upon, to create motor fuels from coal, in effect using coal as a transportation fuel. In the late 1970s and early 1980s, when oil prices in today's dollars were above $50 a barrel, a lot of research was done in this area. In 1980, H. Hillel, a researcher for Germany's energy department, noted: "Proven coal-conversion processes exist which justify immediate use on an industrial scale. The cost of producing motor fuels . . . [is] in range of profitability, given the present price of oil." That was more than a generation ago. Of course, one reason we are still waiting for a massive coal-to-oil effort is that oil prices have come down sharply in real terms since 1980. But as oil prices rise, such an effort will have genuine potential. If we were willing to make the necessary investment in coal conversion, we'd have a real shot at reducing our dependence on oil far more quickly than might be imagined. In chapter 13, in fact, we recommend as a highly interesting speculation a small company involved in just this technology in China.

But what about the argument we made last chapter that the likely environmental costs of using coal are simply too great to ignore, however desperate we are for energy? We're not backing away from that. One point is that converting coal to motor oil isn't the same as burning it directly and has a lesser environmental impact—though it doesn't sidestep environmental issues completely, since it results in carbon by-products that must be disposed of. And, of course, it still commits us to mining coal, with all the environmental degradation and health risks to miners involved. But more relevant, under our proposal we wouldn't be upping the amount of coal we mine—we'd simply be redirecting our use of it.

Above we noted that there are technologies for making coal cleaner—the idea is to separate coal into a clean-burning gas, with carbon dioxide as the by-product. One prototype plant involved in such an effort is the Wabash River Coal Gasification Repowering Project in West Terre Haute, Indiana. Right now, though, the costs are prohibitive, far higher than for wind and for regular coal plants. In fact, to convert the U.S. into a clean coal-generating economy would be twice as expensive as converting our electrical grid to wind. In addition, there are unsolved problems of what to do with the carbon dioxide by-product. Research into this process should continue, but wind energy clearly seems to be where we should put the bulk of our resources for now.

One other alternative energy whose potential would zoom once we developed abundant wind energy is rechargeable batteries, used in electric cars. Electric cars have been touted as a way of cutting emissions from conventional internal combustion engines. But since their electricity comes from fossil fuels, right now they don't represent much in the way of progress. If wind were to become more important in generating electricity—and contingent upon improvements in battery technology—electric cars would be running on renewable energy.

So we think it is essential and urgent to begin a concerted national effort to convert the bulk of our electrical grid to wind energy. To build the wind towers needed to put a large portion of our electric grid on wind energy would be very expensive. It's likely to cost at least half a trillion dollars. This is an enormous sum. But rising oil and gas prices are going to force us to do something, and we are going to be pushed against the wall on energy much sooner than is commonly thought. This is a doable and, we think, reasonable approach.

Moreover, the creation of a vast new industry has a lot of positive aspects as well. It generates a lot of new businesses and a lot of new jobs. It should prove to be a tremendous growth industry. And it will buy us time to work to develop an economy based entirely on renewable fuels, something that if we can accomplish would bring about almost inexpressible benefits.

Obviously, to make any progress on the alternative energy front, money is key. We will have to spend a lot. But a lot of the technology is either in place or getting close. As oil prices rise, efforts to develop and employ alternative energies will intensify. That's part of the capitalistic system. Government also will start to glom on to the issue in a much bigger way. Let's hope that it does so in an intelligent and farsighted way, and sooner rather than later. Wind is a natural place to start.

Key Points:

- ◆ Clean renewable energies account for only one-tenth of one percent of energy usage in the U.S. That level will have to increase exponentially.
- ◆ Solar cells are decades away from being efficient enough to meet our energy needs. Until we can use the sun to crack water, solar energy's potential will remain limited.
- ◆ Despite the hype, cars that run on hydrogen fuel cells aren't on the horizon anytime soon.
- ◆ The technology for wind energy is a lot further along, and costs have come down dramatically.

◆ We propose a massive government-assisted investment in wind energy. Wind could then generate the electricity that now comes from coal. The coal thereby freed up could be converted to motor fuels, replacing the oil now used to run cars.

CHAPTER 8

Government Spending

In the preceding chapters we've described three of what we think will be the most salient and unshakable features of the next decade and beyond. These are the bedrock realities we're stuck with, and they will force our economy and financial markets to travel along certain narrow and very tricky paths. To quickly recap, the first is the geological fact that we are nearing the point where we can't increase oil production. The second is the high level of consumer debt, which makes it essential to keep economic growth on track, in part by keeping home prices strong. The third, the logical outgrowth of the first two, is the need to develop alternative energies so that we have a way to fuel economic growth as oil supplies become less and less adequate.

There is one other trend that will dovetail in perfect harmony with these three and that to some degree stems from them. It is increasingly high levels of government spending, an inherently inflationary factor. In coming years, regardless of whether Democrats or Republicans are in power, government spending is set to soar. This will be the icing on the inflationary cake.

Below we very briefly sketch why rising government spending is both inflationary and inevitable. And when we say briefly, we mean it. It makes sense to go into more detail in later chapters when we make a case for particular investments. Still, we feel that the upcoming rise in government spending is a significant and overlooked inflationary development that clearly deserves a chapter of its own.

And like so much else, it all winds up at oil's door. A big part of the increase in government spending, though certainly not all, will be aimed at combating the effects of the oil squeeze. Once again, looking at the future through the prism of oil yields some intriguing results with strong investment implications.

Spending, Not Deficits

Note that what we're talking about here is government *spending,* as distinct from government deficits. Government spending, and not deficits, is what contributes to inflation. That's because inflation is caused by the demand for goods increasing relative to supply. To the extent that the government spends money, it is raising the demand for goods and thereby adding to inflationary pressures.

The deficit is the difference between government spending and government tax revenues. To understand why deficits are less inflationary than spending, imagine that the government decides to cut taxes to zero but doesn't change how much it is spending. Clearly, deficits will rise by whatever the taxes would have added up to. But this doesn't necessarily mean that resources in the economy will be stretched by comparable amounts. Why? Because people might save some or all of the money they formerly paid in taxes, rather than spending it right away. The deficit, therefore, doesn't automatically increase demand the way government spending does.

When deficits rise, you can only guess to what degree they might nudge inflation higher. But when it comes to government outlays, it's a hard-and-fast reality that whatever the government is spending is vying with other spending for the economy's resources.

It's true that some government spending is productive; that is, it's not just a drag on the economy—it can increase productivity, for instance, by funding improvements in infrastructure or even through supporting education, and it can generate economic growth. Still, government spending is invariably less productive than corporate spending, because corporations, flogged along by profit incentives, are highly motivated to make sure they use the money they spend productively. With government spending, a portion always tends to be nonproductive.

The historical record bears out our point about the inflationary impact of government spending. Whenever government spending has risen at a rapid rate, becoming a significant portion of the gross domestic product, inflation has taken off. Since the beginning of the twentieth century, a threefold gain in government outlays over a ten-year period resulting in government spending that is more than 20 percent of the GDP has led to inflation that ultimately reached double-digit levels. As figure 8a,

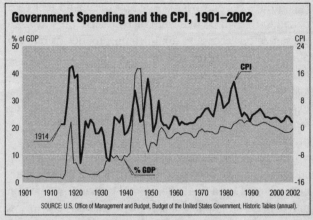

Government Spending and the CPI, 1901–2002

SOURCE: U.S. Office of Management and Budget, Budget of the United States Government, Historic Tables (annual).

Figure 8a

"Government Spending and the CPI, 1901–2002," shows, there have been three periods that met these criteria: World War I, World War II, and the mid-1960s through the early 1980s.

The Defense/Energy Interlock

It's amazing how oil insinuates itself everywhere, oozing into every economic nook and cranny. Government spending is no exception. In coming years, the energy crisis will begin to influence government policies in a variety of ways. One of the most significant will be to force a big rise in defense spending, which will account for a major proportion of overall spending increases.

How does oil relate to a surge in defense spending? Because the major purpose of building up our military capabilities beyond what we otherwise would feel compelled to do will be to ensure that we have the muscle to guarantee our access to diminishing supplies of oil, supplies that much of the rest of the world will be trying to get its hands on as well. We go into more detail about this and other factors relating to increased defense spending in chapter 14, where we recommend the best-situated defense companies as core investments for just about every investor.

Interestingly, because of our focus on forthcoming oil shortages, we've been predicting rises in defense-related expenditures for a while now—and certainly well before our expensive war and subsequent engagement in Iraq were even being debated. But the war in Iraq was right in line with our analysis. You can debate our motives for that war—whether it was purely and simply a belief that Saddam harbored weapons of mass destruction, whether we altruistically wanted to liberate the Iraqi people from a hateful tyranny, whether it was the long-planned opening act in an ideologically motivated push to reshape the whole framework of American foreign policy. But it would be naive to ignore that the context for the war was that Iraq is in an oil-rich part of the world, in fact, the only part of the world with any meaningful long-term excess oil-producing capacity. Whether that fact necessitated going to war is debatable, and maybe the above reasons were really what lay behind our decision. What's not debatable is that we need to have a way to protect our access to Middle Eastern oil, and that to back that up we need a bolstered military capability.

Spending on Alternative Energies

While the oil crisis will force an increase in government spending in one way, through its impact on defense budgets, it will increase it in another way by forcing the government to assume a major role in developing alternative energies. We gave some of the math last chapter, and as we indicated, the numbers are huge. It's anyone's guess what the final figure will be, but expenditures approaching $1 trillion on alternative energies are not inconceivable. It's true that private enterprise will also play a big role, but history shows that when it comes to massive projects— and the development of alternative energies could be as big a project as a major war—the federal government inevitably gets involved.

True, so far no energy legislation has tackled the problem in a meaningful way. But we are still in the early stages of what will be a long-term and intractable crisis. If, as it has been said, hanging wonderfully concentrates the mind, $100 oil should go far toward clarifying our real options.

Social Spending

One reason inflation was so high during the latter 1960s was that President Johnson wanted to fight the war in Vietnam without sacrificing social programs at home, in a formulation that came to be known as "guns and butter." In today's world, there will be comparable pressures on the government to keep spending on social initiatives, which in today's environment come on top of energy as well as military outlays. One reason is that four-letter word, "debt," that overhangs the economy like a sword of Damocles. With the huge consumer debt levels making the economy so vulnerable to any hint of a slowdown, the government will have to get into the business of stimulating the economy with additional spending anytime the situation calls for it. That was the case after September 11, when huge government expenditures were probably a big reason the economy didn't go into a tailspin. You could argue that big increases in expenditures on defense and alternative energies will be enough stimulation to keep the economy out of trouble. That may be true, but it also may be true that at times additional stimulation will be needed.

In any case, demographic realities are going to start coming to a head within the coming decade, and they all will require that the government spend a lot of money. America is getting older. The median age of Americans has been rising since 1970, and it is likely to continue to rise until the middle of this century. The number of older Americans will be rising, which will mean a greater need for health care. And we haven't even touched on Social Security, for which, absent any major changes, expenditures will start to explode next decade as baby boomers begin reaching retirement age. Maybe to compensate for surging spending on energy and defense we'll end up making such compromises as increasing the retirement age or cutting health care in a big way. But keep in mind that politicians will always need to get elected, and the growing number of seniors will include a lot of voters.

This is a brief rundown of why government spending is slated to rise sharply in coming years. It fits hand in glove with the overall picture of a world that will be dominated by the need to find energy and keep the economy growing. And it adds up to one reality: rising inflationary pressures.

Key Points:

- ◆ Government spending is set to soar, which will further add to inflationary pressures.
- ◆ One reason will be big increases in defense spending—spurred in large part by the need to ensure our access to tight oil supplies. A second reason will be the need to spend big on alternative energies.
- ◆ Spending on social programs will also remain high as the population ages.

CHAPTER 9

The Case for Inflation

We've laid out a handful of key trends that we think will assert themselves with increasing ferocity as time goes by—oil shortages, rising oil prices, the need to keep home prices high, and rising government spending. As we've described, they are likely to combine to lead to years of rising inflation punctuated at irregular intervals by briefer periods when deflation looms up as a concern.

Here we wrap up a few final loose ends to explain further why inflation will be so difficult to head off. First we detail the historical relationship between rising energy prices and high inflation. Second, we show that in striving to keep home prices high, the Fed will have to tolerate negative real interest rates—that is, interest rates that are lower than inflation—which in and of themselves are inflationary. And third, we explain why we can't count on productivity gains to keep inflation well behaved. These might seem like a somewhat random assortment of topics, but they all will contribute in important ways to the economic environment, and they all point to higher inflation.

Energy and the CPI

In his book *The Great Wave,* a history of inflation from the thirteenth century to the present, author David Hackett Fischer described trends in England in the seventeenth century as follows: "The price of manufactures also rose at a slower pace than those of food and fuel. Throughout Europe, the slowest rates of increase were for industrial goods which could be produced most easily in larger quantity. In England, the price of food and fuel rose by a factor of six or eight, while industrial commodities merely trebled." And he added: "That pattern of price relatives has appeared in every great wave."

His point: not just in England in the 1600s but on a consistent basis over the last eight hundred years or so, energy prices that were rising more rapidly than prices overall have been the tip-off to the fact that inflation was about to take off in a big way. You might argue that just because something was true for centuries doesn't mean it's true today. Granted. But in modern history, too, sharply rising energy prices have been a key factor in exacerbating inflation overall—even, in some cases, as in the early 1970s through the early 1980s, causing inflation to remain high in times of recession, resulting in stagflation.

If you look back at figure 8a, you can trace the course of inflation, as measured by the annual percentage change in the consumer price index (CPI). The CPI is a basket of goods supposed to represent what a typical consumer spends money on. Different items have different weights—housing, for instance, gets more respect than peanut butter. While the CPI is far from a perfect indicator, it is still the measure of inflation that goes back the furthest. As you can see, since 1914 there have been three periods in which inflation really soared, climbing above the 10 percent level. The first was in World War I, the second in World War II, and the third was from the mid-1970s through the early 1980s. Clearly, the first two eras of high inflation were special cases, in that large-scale wars entail shortages of a multitude of commodities, from oil to foodstuffs to rubber. So we don't think it makes a lot of sense to try to draw conclusions about inflation from times of war.

The third period of sharply rising inflation is a different story. For about a generation after World War II, apart from a brief spike in 1951—again war-related, this time the Korean War—inflation remained tame, rising above 3 percent just once, in 1957. In the stretch between 1959 and 1965, inflation never topped 2 percent. But then, in the late 1960s, propelled by the high government spending we noted earlier, it began to take off, prompting President Nixon to institute wage and price controls in August 1971. These, for a while, tamped down inflation, with the CPI falling from nearly 6 percent in 1970 to a touch over 3 percent in 1972, when the controls were lifted. At that point, with government spending still accelerating, the CPI took off again.

Now look at figure 9a, "Energy and Inflation," which traces inflation in conjunction with the price history of a broad-based measure of energy prices, including oil, coal, and natural gas. Between 1966 and 1972, energy-

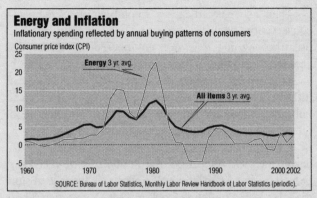

Energy and Inflation

Inflationary spending reflected by annual buying patterns of consumers

Consumer price index (CPI)

SOURCE: Bureau of Labor Statistics, Monthly Labor Review Handbook of Labor Statistics (periodic).

Figure 9a

related inflation remained below overall inflation. Clearly, at this point energy prices weren't the engine of overall inflation; they were going along for the ride. In 1973, however, energy-related inflation really surged, pushing energy inflation above overall inflation. When the oil embargo kicked in, overall inflation topped double digits for the first time since World War II. Even the recession of 1973–75 didn't check price rises much, as energy prices continued to move up. In 1976 energy prices were still rising at an annualized rate of nearly 10 percent. While overall inflation moderated, it still remained close to 6 percent—more than twice the average of the 1955–70 period.

By 1979, with energy prices remaining sharply uptrended, overall inflation topped the 11 percent level reached in 1974, and in 1980 the CPI at just below 14 percent was nearly as high as during World War II. Despite the recession in 1981, the CPI stayed in double digits.

Then, as we noted earlier, a combination of factors—conservation, nuclear energy, and additional oil supplies, plus the ongoing recession engineered largely by Fed chairman Volcker's anti-inflation crusade—finally checked energy inflation in the early 1980s. And overall inflation began a generation-long downtrend, interrupted only by the first Persian Gulf War, which kicked up both oil inflation and overall inflation to over 5 percent.

Energy prices and the CPI remained well under control during the 1992–99 period. But in 2000, as we noted earlier, oil prices spiked, and inflation even in the face of slow growth and the tumbling stock market rose to over 3 percent, its highest level in almost a decade. In 2000–2003, as energy prices remained above the levels that had prevailed for most of the 1990s, overall inflation remained elevated as well, despite another recession and overall very slow growth, indeed, growth slow enough to sharply check the growth in demand for oil.

The historical record is clear: once energy prices start to rise sharply, they will make it almost impossible to stop inflation, even during a recession. And this is perfectly logical, since energy is an integral part of virtually every aspect of the economy. Even the computer on which these words are being typed depends on electricity. Higher energy prices force businesses to charge more for goods, which in turn forces workers to demand higher wages, further adding to the cost of goods. And so it goes. When energy inflation gets going, it tends to pull everything along in its wake.

The Inflation Tipping Point

Let's look at how things are likely to unfold over the next few years. Oil prices will rise; there may be periods of relative weakness, but the overall trend will be sharply up. With oil prices rising faster than those for other goods and services, oil and energy will become much more crucial components of the entire economy. This is a significant concept that's important to understand. Compare energy with platinum, for instance. Platinum, in addition to being prized in jewelry, is also a vital industrial metal. A whole host of products that are key to this economy, from cars to computers, need platinum or related metals, and in many cases there are no substitutes. Still, while essential, platinum accounts for just a small amount of the overall cost of these goods. Even if platinum were to double in price, it would have very little effect on how much it costs to make these products.

But suppose platinum kept doubling in price. Eventually it would become a key part of the overall cost of producing cars and computers and many other goods. Suddenly the trend in platinum prices would be of prime importance to our economy's health.

Platinum is not likely to take on such a role. But energy has and will. In the early 1980s energy expenditures accounted for more than 13 percent of economic activity. As oil prices receded, so did oil's importance in the cost of goods and services. By the late 1990s energy expenditures accounted for just 6 percent of GDP. This relatively small proportion is one reason that the big rise in energy prices in the early 2000s did not dramatically boost overall inflation. (Another and equally important reason was the fall in overall demand resulting from the market crash and the subsequent recession in 2001.)

Economies are complex, and there is no magic level after which rises in energy prices suddenly have a determinative impact on the whole economy, no single number to watch out for. All we can say is that we are close to that point. In the mid- and late 1970s, energy expenditures represented about 10 percent of overall economic activity and clearly were a driving force in the inflation that gripped those years. Even as early as 1973–75, when energy expenditures were about 8 percent of economic activity, energy still was a significant driver of the ultra-high inflation that prevailed during those years. Though the data that make up these numbers are complex and are usually reported with multiyear lags, the best guess is that today energy expenditures are more than 7 percent of the economy. We don't have a lot of leeway.

Nominal vs. Real Rates

Earlier, in chapter 5, we explored the inflationary impact of high levels of consumer debt. To recap, high consumer debt makes the economy more leveraged and makes it essential to keep home prices high, which means economic growth can't be allowed to falter, effectively handcuffing the Fed. Strong economic growth will mean no letup in demand for oil, which means there will be no way to ward off ever higher oil prices. And as we just outlined, higher oil prices historically mean higher overall inflation. We also explained why in inflationary times, real assets such as homes are desirable investments—it's because their value keeps up with inflation.

You might think, or hope, that we've said all we need to say on this subject. But actually, to really understand why the Fed won't be able to mount much resistance to inflation, it is worth going a step further to look at the role of interest rates in more detail. In particular, it is necessary to focus on the difference between real interest rates and nominal rates.

When a bank advertises that it pays, say, 3 percent on a savings account, that's the nominal rate. Similarly, when you borrow money, for example, by taking out a mortgage, and you're told that you're being charged interest of 3 percent, that's the nominal rate. Nominal rates are absolute rates. What they don't do is tell you whether you're getting a good deal.

Real interest rates tell you what you're earning, or being charged, in relation to inflation. They are nominal rates minus inflation, and they are a more meaningful measure than nominal rates. Suppose you're still earning 3 percent on the money in your savings account but inflation is 5 percent. This means that you're losing ground—after a year, your money, including the interest it has earned, will buy less than it would have at the start of the year. But if inflation is only 1 percent, then your buying power is growing.

Now think about some of the implications of real rates being negative, meaning that inflation is greater than the nominal interest rate. For one thing, under these circumstances, it makes no sense to put money in a savings account—you're losing out from day one.

On the other hand, when real rates are negative, it makes a lot of sense to borrow money to buy a home, which is a hard asset that doesn't get consumed and whose value rises with inflation. Assume inflation is 10 percent and nominal interest rates are 5 percent, meaning real interest rates are negative 5 percent. If you take out a mortgage to buy a house, you're paying 5 percent interest on your loan, while your house is going up in value by 10 percent. You're making money from day one.

Clearly, negative real interest rates are manna for the real estate market. Similarly, they are an incentive to buy hard assets in general, in particular gold and other precious metals and anything else whose long-term value keeps up with inflation.

Negative real interest rates are inherently inflationary and create the potential for a vicious inflationary spiral. Once investors see that they get an immediate return when they buy hard assets, the demand for such assets rises, which pushes their prices higher—i.e., leads to greater inflation. Concomitant rises in the price of homes reinforce the need for higher wages. Moreover, higher home prices also give homeowners greater access to credit and thus help to increase spending on other goods, which puts further pressure on prices.

So you can easily see why there is a strong inverse relationship between real interest rates and inflation. Figure 9b, "Inverse Relationship," shows this clearly, using the yield on 91-day T-bills as our interest rate measure and the CPI's 12-month rate of change as our measure of inflation. Between 1972 and mid-2003, when real rates were negative, inflation averaged nearly 8 percent, more than twice as high as when real rates were positive. And when real rates were strongly negative, below minus 1, inflation really went on a tear, averaging nearly 10 percent.

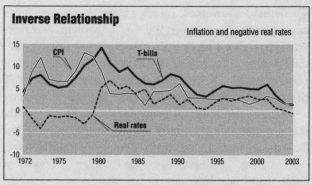

Figure 9b

An Impotent Fed

When real rates are negative, and inflation is heating up, the Federal Reserve, which controls T-bill yields, theoretically can simply raise rates enough that real rates turn positive. Presto—you've pulled the carpet out from under all those inflationary impulses, scattering them to the winds. This is exactly what Paul Volcker did in the early 1980s. When Volcker became Fed chairman in 1979, inflation was in the low double digits, nominal rates were a few percentage points below that, and thus real interest rates were heavily negative, below minus 1. It took Volcker a little time to get his act together and formulate a plan, and by early 1980 inflation had risen further, approaching the worst levels of World War II. Real rates had continued to deteriorate and were below minus 7. Gold was flying, and the vicious circle of which we spoke, in which high inflation begets even higher inflation, was in full swing.

Then Volcker acted, raising short-term interest rates dramatically. By the time 1982 rolled around, real rates had swung all the way back to about plus 4. They continued to climb into 1983, after which, with the exception of a few months in 1993, they remained positive until 2003. During this long stretch of positive real rates, inflation went from terrorizing the economy to a no-show. Its absence set the stage for one of the greatest periods ever for financial assets and one of the worst ever for real assets such as gold, which is perhaps the ultimate barometer of inflationary pressures.

In today's far more debt-heavy world, however, a little more than a generation after Volcker worked his disappearing act on inflation, the same bag of tricks won't be available. Volcker tamed inflation by acting to turn real rates sharply positive. Today, turning real rates sharply positive would be a disaster because, as we explained above, the real estate market thrives on negative real rates. And as we can't stress too often, because of the huge overhang of consumer debt, the primary concern of policymakers has to be keeping home prices high. Negative real rates are likely to become the norm, and they will be yet another inflationary force.

If inflation really starts to soar, and real rates turn sharply negative, we're not saying the Fed won't try at all to cool things off. But it won't be able to proceed with the same decisiveness and single-mindedness as Volcker did in his day. At most it will act gingerly, raising rates with great trepidation and a probable willingness to back off. The memories of 2000–2002, when rising home prices were all that saved the economy from total collapse, will surely linger, and rightfully so.

We know we've plied you with a lot of information. If it seems too confusing, just realize that from an investor's point of view, there are two key points. The first is that inflation favors hard assets. And the second is that even in the absence of significant inflation, negative real interest rates also favor hard assets—though it's also true that negative real interest rates tend to foster inflation. In the years ahead, the odds are high that we'll see high and rising inflation. But they are even higher that we will see negative real interest rates. And in terms of investments, this leads to the same place.

The Productivity Myth

We know we haven't painted a very pretty picture of what the next decade or longer most likely holds in store. Stubborn inflation underpinned by the unholy trio of rising energy prices, the need to keep home prices high through negative real rates, and sharp increases in government spending—surely there must be some ray of hope we have overlooked, some knight on a white horse that can ride to our rescue. Yes—there it is—it's rising productivity! No, wait, sorry, we were dreaming.

Productivity has been a much-invoked and much-misunderstood concept in recent years. Its most general definition is the ability to do more with less, and it's easy to see why rising productivity can counteract inflation. When a business can make the same product for less money, it can charge consumers less. Technological innovations are usually the force behind productivity increases because they increase efficiency. They enable a business to use fewer workers to turn out the same number of products or, in the service arena, to deal with the same number of customers. Or, of course, what amounts to the same thing, the same number of workers may turn out more products. That's the basic idea, and under this definition, productivity gains can be measured in a pretty straightforward way.

There's a corollary to this definition of productivity, however, that muddies the waters somewhat. Theoretically if a business makes a better product for which it charges the same price or even a somewhat higher price, you also could say productivity has increased, because you're getting more for your money. Let's give a really simple example. Suppose it used to cost $35 to buy an old-fashioned coffee percolator. Now suppose that you can't find these anymore and instead you have to spend $45 to buy an up-to-the-minute coffeemaker that has all kinds of wonderful features. You can program it the night before to have your coffee ready at exactly 7 A.M., you can adjust the water flow depending on how many cups you're making, and so on. It does all but solve the morning crossword puzzle for you. But it costs $10 more than your old coffeemaker, and in the end, you're still drinking more or less the same coffee. How do you assess productivity in this instance, and is the impact inflationary or not?

We'll give you a hint where we're headed. In the late 1990s, based on how they interpreted other situations, the government functionaries in charge of calculating inflation probably favored the view that the new coffeemaker did *not* represent inflation, because consumers supposedly were getting more value for their dollars. At least, this was their approach to a whole lot of decisions that revolved around issues of qualitative improvements. And we think they were mostly dead wrong.

Why is this relevant to whether the years ahead will be inflationary? Because if, as we believe, since the mid-1990s productivity gains involving presumed qualitative improvements, particularly in technology, were overstated, it means inflation has been higher than generally reported. Productivity has not been the miraculous economic fix-it that many people have thought. Even more important, it won't fulfill that role in the future. We can't count on productivity to slay the inflation monster, which knocks off one of the few hopes of keeping inflation low.

The McKinsey Report

It's worth looking at productivity in more detail, because it relates so closely to prospects for inflation and because it will continue to be widely misunderstood and misinterpreted. Let's return to the 1990s, which were generally considered a productivity miracle. In particular, between 1995 and 2000 productivity is widely thought to have grown at a relatively rapid 2.5 percent clip, compared with just 1.4 percent in the years between 1974 and 2000.

The most comprehensive study of productivity in 1995–2000 was done by the big consulting firm McKinsey & Co. Its report noted: "Nearly all of the post-1995 productivity jump can be explained by the performance of just six economic sectors: retail, wholesale, securities, telecom, semiconductors, and computer manufacturing. The other 70 percent of the economy contributed a mix of small productivity gains and losses that offset each other."

Let's see what really was going on in the sectors McKinsey singled out, starting with information technology, which encompasses computers, semiconductors, and telecommunications. The basic assumption in ascribing productivity gains in this area was that, like our coffeemaker, computers got "better"—faster, with more complex capabilities—and also that we got more efficient at making them. The bean counters in government bureaus in charge of calculating productivity and inflation statistics were dazzled by these accomplishments. They assumed that if a newer computer performs its calculations twice as fast as an earlier model, and its price stays the same, then productivity in the computer industry has jumped 100 percent. And with computers an important segment of our economy, this boosted overall productivity figures. It's a major reason why even as food and transportation and other prices were rising, the government kept insisting that inflation was low.

But the bean counters were missing something critical, which is that unlike houses, food, and other essentials, computers are not significant in their own right—they are important only as a tool that enables us to do something else. The important question that no one asked was whether we really were doing more with these new, improved machines. Does faster really mean better? We'd argue no. And if that's true, then there is something very self-referential in the argument that creating a jazzed-up tool that doesn't really accomplish much more than the previous model is somehow a great economic boon.

An illuminating example that supports the view that faster computers aren't better computers comes from the world of chess. In 1997, Garry Kasparov, then the world's chess champion, played a six-game chess match against the world's best chess-playing computer, dubbed "Deeper Blue." Kasparov seemingly lost—out of six games, he lost two, won one, and drew three. His defeat was heralded as the start of a new age in which artificial intelligence was beginning to outrun human intelligence in a wide range of problem-solving activities.

As avid chess aficionados, we argued in our last book that the result had been grossly misinterpreted and that, in fact, Kasparov's defeat resulted from a sudden loss of nerve—among other uncharacteristic mistakes, he resigned from a drawn position—rather than from an inability to match wits with a computer.

Now flash forward six years. Computers in 2003 have far faster chips and more memory than in 1997—they are quantitatively better in every way. Not true for Kasparov, who has been supplanted as world chess champion and is clearly past his prime. Human brain cells, alas, don't keep getting faster as their owner gets older. If quantitative improvements in computers were truly meaningful, the rematch that took place in 2003 should have been a spectacular mismatch. The computer should have trounced Kasparov thoroughly.

That's not what happened. The 2003 rematch ended in a draw. Moreover, once again, Kasparov, because of a clear failure of nerve, underplayed his abilities. Among other things, in the final game he offered a draw even though his position was clearly stronger than the computer's and he stood to win. He confessed later that his goal wasn't so much to win as simply not to lose.

Faster computers are better computers? Tell this to a chess junkie. It just isn't true.

Similarly, there also is less than meets the eye to the notion that productivity increased sharply in the securities industry, one of the other areas the McKinsey report highlighted. The market meltdown at the start of the decade shows that there is more to productivity in this area than the ability to trade more stocks and bonds with fewer people.

Another big sector highlighted in the McKinsey report was retail, and here there were some bona fide gains in productivity in the second part of the 1990s. As retailers cut back on the number of workers they needed to manage and handle merchandise, cost savings were passed on to consumers. But the impact of these gains will be limited. As the McKinsey study pointed out, one reason for the productivity increases in retailing was the looming presence of Wal-Mart, which forced competitors to streamline their operations. In other words, one admittedly extraordinary company was the driving force behind many of the gains in retail productivity. And while Wal-Mart will surely remain a retailing giant, it's unlikely that the outsized gains it forced upon the overall retail industry in the late 1990s will continue to be repeated.

In the wholesale arena, one sector that was rated as having experienced a large jump in productivity was pharmaceuticals. This may surprise you, given that health care costs, spurred in large part by rising drug expenditures, have been rising far faster than inflation in general. The thinking, though, is that in the second half of the 1990s a whole array of wondrous new drugs came on the market, with far superior abilities to fight or prevent disease. So even though consumers were paying more, they were getting better-quality drugs that were more than worth the extra money. Again, it's the coffeemaker syndrome, the idea that a "better" product, even if it costs more, is noninflationary.

But in the drug arena, too, this is highly debatable. One such blockbuster drug was Lipitor, a cholesterol-lowering medication made by Pfizer. It became an instant hit, widely prescribed by cardiologists. You can make a good case, however, that Lipitor is no more effective than extra doses of niacin—an inexpensive vitamin doctors have long used with heart patients—in concert with diet and exercise. The same goes for a lot of other new drugs.

In general, once you put government statisticians in charge of assigning a value to quality, virtually anything goes. So-called quality improvements have been instrumental in the supposed productivity miracles of the 1990s, and therefore in the relatively slow inflation rate believed to have prevailed. If you believe, as we do, that the improvements in quality are for the most part either illusory or minimal, it means that inflation has been vastly understated. We simply don't buy the official version.

One compelling piece of evidence that we are right revolves around supposed increases in living standards. According to any of the measures used by the government, standards of living in the U.S. have improved enormously since 1970. Roughly speaking, increased standards of living result from increased productivity—consumers are getting more for less money or better for less or the same or just a little more money. But this just doesn't jibe with the fact that in 1970 only 30 percent of married women with children under the age of six worked outside the home, while in 2000 the percentage was 63 percent. Changing social mores may explain some of the difference, but not most of it. The reality is that for a growing number of American families, two incomes are required to provide the necessities that one income could provide in 1970. This is just not consistent with rising living standards.

In short, a belief that rising productivity has kept and will keep inflation tame is a bit like believing in Santa Claus. It's a nice idea, but the weight of the evidence is against it.

◆ ◆ ◆

There you have our case for inflation, all nine chapters of it. Together, all the trends we've been describing are likely to cause inflation to take hold with the tenacity of a demented terrier that has sunk its teeth into some unfortunate prey. Investors need to start preparing for a turbulent new era of rising prices.

And this requires a whole new mindset. For when it comes to investing, inflation is not just another economic statistic. It's what determines the whole investment environment, sets the investment agenda. Choosing investments without regard to prospects for inflation is like taking a vacation without bothering to learn about the climate of the place you will be visiting. You can have fun in any climate—as long as you are prepared. But if you go to Vail in January and bring only shorts and bathing suits, or if you go to St. Thomas and pack only ski parkas, you're going to miss all the fun. The same goes for investing. If you shop for investments that match the inflationary climate, you can win big whether inflation is raging or a no-show. Ignore inflation, though, and you may find that you don't have enough money to cover the entrance fee to Six Flags Great Adventure, much less to let you luxuriate for a week in Vail or the Caribbean.

This brings us to part II, our investment section, in which we tell you exactly how inflation will turn the investment climate of the 1990s on its head and what steps you should be taking to inflation-proof your portfolio.

Key Points:

◆ Over the past eight hundred years, energy prices rising faster than other prices have led to runaway inflation.

◆ When real interest rates are negative, meaning inflation is higher than nominal rates, it pays to put money into real assets, including homes and precious metals. To keep home prices strong, the Fed will likely keep real rates negative for much of the time in coming years—and negative real rates are inherently inflationary.

◆ Productivity gains have been less than generally thought and won't be the answer to rising inflation.

Part II

Making Money

Now we get to the payoff—how to cash in on the trends we've described. In this section we cover one by one the groups that will be the chief beneficiaries of inflation and negative real interest rates, as well as of the final explosive stage of oil's dominance. We also warn you about what to avoid, tell you how to hedge your bets by including deflationary plays, and explain how to use our oil indicator to adjust whenever the economic background alters.

We ended the last section with a general survey of all the factors that are likely to lead to inflation. This section starts with another look at inflation—this time from a direct stock market perspective. We tell you how high and rising inflation affects P/Es and what essential characteristics any stock needs to flourish in inflationary times.

Getting and sticking with the program is crucial. When inflation rides high, you can't win with any old stock, no matter how much the company has going for it, if it can't grow rapidly enough to outpace inflation. In times of inflation, investors need to be both more disciplined and more vigilant. We explain how to go about achieving these winning ways.

We've included all the groups we can think of that are almost sure to be standouts in coming years. But we hope that we've also given you the tools to enable you to recognize on your own additional investments that might emerge and meet our investing criteria. It's the old teach-a-person-to-fish idea. Still, if you do nothing more than stick with the ideas we present here, you should outperform the huge majority of investors, amateurs and even pros alike.

It's worth reiterating one point. Inflation is the most likely scenario. But even in our second most likely scenario, lower inflation in the context of negative real interest rates, the same assortment of investments, with just a few small exceptions, is still your best bet. In other words, whether we have bona fide inflation, as we expect, or just an inflationlike environment, the investment choices are the same.

To help you pull together the advice we've given you, our final chapter (save for the epilogue) offers model portfolios in which we suggest reasonable allocations among all the various investments that we think investors should own. There is no one-size-fits-all portfolio, of course, but these come close. We offer two portfolios, one for when our oil indicator is positive, one for when it turns negative.

One final caution: if when this book comes out, energy prices have retreated and inflation is low, don't assume that all our advice is canceled—not unless some hugely unexpected event has occurred, such as some god descending to earth to offer the gift of limitless nonpolluting energy. But absent that or the discovery of a new and entirely unknown North Sea that has been languishing in Brigadoon for all these many years, oil prices will move higher and higher, and inflation will too. Our projections are for the long term—as you've no doubt noticed, we constantly refer to time frames of the next decade or longer—and it may take a little while longer before the trends we've described emerge clearly into daylight. Don't lose sight of the big picture.

CHAPTER 10

The ABCs of Inflation Investing

Inflationary times, as we noted last chapter, lead to an investment climate radically different from that in low-inflation periods such as the 1990s. It's one that's far less balmy, far more challenging, and very unforgiving. Actually, it's less Vail and more Mount Everest. There are fewer right choices and a lot more wrong choices, and you need continual vigilance. If you could choose what type of investment climate to be in, you'd never pick inflation, not unless you have severe masochistic tendencies. Unfortunately, unlike planning a vacation, you don't have any say in the matter. You just have to accept the hand you're dealt, and the key is to recognize it for what it is and adapt so as to make the most of it. And that's where we are today, for as we've been stressing, the odds strongly favor that we're entering a new inflationary, or at the very least inflationary-like, period.

There are two cardinal rules for successful investing when inflation heats up. The first is to buy only those companies whose earnings growth is strong enough to compensate for the fact that—as we'll explain shortly—P/E ratios will be downgraded across the board. This leads you inexorably to limit your investments primarily to a fairly restricted group of companies that are leveraged to inflation—that actually benefit from inflation. In particular, it leads you to focus largely on companies whose businesses are based directly on real assets, for those companies' earnings are leveraged to price increases in the underlying commodities. Somewhat counterintuitively, this rule also sometimes leads you to shun some of the most solid and best-managed companies around, because in inflationary times, great companies can be lousy stocks.

The second rule is to abandon a policy of buy and hold. When inflation is in the driver's seat, the irony is that you have to be ever alert for inflation's alter ego—deflationary fears. You need to be able to shift among two sets of investments, those geared primarily to inflation and those better suited to deflation. If, using our oil indicator as your guide, you can be flexible and proactive, you will see your winnings really start to mount.

In the following chapters we tell you exactly how to apply these two cardinal rules—what to buy, group by group and stock by stock, and how to maximize your gains through careful market timing. Here we explain why and how inflation so thoroughly transforms the investment environment.

It's All Relative

If your teenaged son goes to the mall and spends $50, did he spend a lot or a little? Well, if he returns with two new pairs of pants for school, three dress shirts and four good-quality T-shirts, a bunch of underwear, and a pair of sneakers, we'd say he spent very little. (We'd also say it probably wasn't our son.) If he comes back with two pairs of designer socks and a tattoo (and trust us, this is possible), we'd say he spent a lot. Obviously, the $50 by itself tells you nothing. It means something only in relation to something else—the total amount purchased, which gives you a way of making some assessment of value.

When it comes to money matters in general, all of us understand, at least to some extent, why absolute numbers may have meaning only in relation to inflation. If your boss gives you a 25 percent raise, is that good or bad? It depends on inflation. If inflation is running at 5 percent, it's great. If inflation is at 40 percent, though, look for a better boss. Obviously, this is akin to the difference between nominal and real interest rates that we discussed last chapter.

While this is a basic concept, sometimes even the most experienced stock market observers can overlook it. The always entertaining and perennially bearish *Barron's* editor, Alan Abelson, writing in March 2003, compared the market at the end of the 2003 invasion of Iraq with the market at the end of the 1991 Gulf War and concluded that the 2003 market was in a far more precarious position. One of his chief arguments was that stocks in 1991 were cheaper than in 2003, and one piece of evidence he cited was that dividend yields in 1991 were well above 3 percent, compared to yields below 2 percent in 2003. So in 1991 your income from stocks would be more than 50 percent greater than twelve years later. The problem with this reasoning, though, was that in 1991 inflation was over 5 percent, while in 2003 it was under 3 percent. Now, would you rather earn 3 percent when inflation is 5 percent, or 2 percent when inflation is around 2 percent? Obviously the latter is a better deal. Thus, stocks in 2003 were clearly more reasonably valued than in 1991, undermining Abelson's bearish conclusion.

These are some of the simpler and more obvious ways in which inflation acts as a reference point that guides how you view a particular number. But inflation has a somewhat more complex role when it comes to one particular key stock market variable—the price/earnings ratio, or P/E. We noted above that during times of high inflation, P/Es get lowered across the whole spectrum of stocks. Another way of saying this is that when inflation is high, investors view stocks' earnings with a lot more skepticism than when inflation is low. They literally place a lower value on those earnings. And this has wide repercussions on stock market returns in general in inflationary versus noninflationary times and on what kinds of stocks in particular do well during periods of inflation. Understanding how all this plays out will be your ticket to successful investing in the high-inflation years that lie ahead.

Inflation and P/Es

Dividend yields, which Abelson used above as his measure of the market's cheapness, are just one way to assess the value of stocks. They're actually not that satisfactory a method, because a company might have a lot of excellent reasons for keeping dividends low, such as the desire to use profits to invest in ventures that will help the company grow faster. So unless you're interested only in dividend income, differences in dividend yields don't necessarily tell you that much.

A more useful approach is to look at P/Es. P/Es are the price of a stock divided by earnings, and they tell you how investors in the aggregate value a stock. Or to be more specific, they tell you what investors think about a company's earnings—whether they think they're worth paying a lot or a little for.

Suppose you have two companies, each of which has earned $2 a share over the past year. One has a stock price of 20 per share, giving it a P/E of 10. The other has a stock price of 30, giving it a P/E of 15. What this tells you is that investors think the prospects going forward are much better for the second company—that it seems to be better situated to grow.

Earnings growth is what makes investors lick their chops and rub their hands together, and it's what drives P/Es. It's true that to some extent P/Es also factor in relative degrees of risk in a company—that is, how much you can count on its growth. But in general, the faster a company can grow, the better. In the heyday of the wild 1990s investors avidly followed every earnings report, and if earnings for a particular stock came in one penny below expectations, they sold in droves, while if earnings were one penny above expectations, they sent the stock soaring. This was carrying an interest in earnings growth to an unhealthy extreme, and it brought a whole host of problems in its wake. But it shows just how central earnings growth is to how investors feel about a stock.

One final thing you need to understand is that earnings growth has two components: a real component and an inflationary one. Real growth comes from all the things a company does to become a better, more successful company—things like opening more stores, moving into new markets, introducing innovative new products, taking market share away from competitors, cutting costs, and increasing efficiencies. Inflationary growth is a lot more passive: it comes from charging higher prices to cover the higher costs that come with rising inflation. (You might wonder whether the higher costs and higher prices wouldn't cancel each other out. They don't, because it is in the nature of inflation that prices rise faster than costs.) The inflationary portion of growth is roughly equivalent to the general inflation rate, but there are variations depending on the particular company and industry. That is, some companies have an easier time passing on higher costs, while others might need to proceed more slowly.

With all this as background, you are now in a position to understand why it is that P/Es start to sink as inflation rises. It's because investors generally place very little value on the inflationary part of growth. And as inflation rises, that portion becomes an ever more significant portion of overall growth.

Let's see how this works with a simple example. Suppose you have zero inflation, and a company has real growth of 5 percent a year. Let's say its earnings are $2 and its price is 20, giving it a P/E of 10. After one year, its earnings have risen to $2.10—5 percent of $2 is $.10, which you add to the $2—and, with a P/E of 10, its price is 21. Its price, in other words, has gone up $1, representing a 5 percent gain—a gain exactly in line with its real growth. You've made 5 percent, and it's a real gain, because there is no inflation to eat into it.

Now suppose that everything else starts out the same as before except that inflation is at 10 percent. This means that the company's earnings are now growing by 15 percent—the result of adding 10 percent inflationary growth to 5 percent real growth. But investors don't place much value on the inflationary portion of growth—they're still interested only in the 5 percent that represents real growth. As a result, the stock price will still rise to 21, as it did before, and no higher. But with earnings now coming in at $2.30, not $2.10—a 15 percent increase versus a 5 percent increase—the P/E has to come down, in order to keep the price at 21. The basic formula for determining the price of a stock is earnings times P/E, and if the earnings part rises, the P/E ratio needs to drop to get the same answer. You'll find, if you do the math—which involves dividing the price, 21, by the earnings, $2.30—that the new P/E is a little above 9. Meanwhile, you've earned 5 percent—the gain from the stock rising from 20 to 21. But with inflation at 10 percent, you're a loser, because you've lost ground to inflation.

Obviously this is a purely hypothetical example, and in real life everything would be unlikely to work out with such military precision. But the general point is true:

investors are relatively unimpressed by the inflationary portion of growth, and this means that P/Es move lower during inflationary times. It also tells you why under these conditions your investments are likely to lose ground to inflation—it's because stock prices are more apt to keep up with gains in real earnings than with the higher nominal earnings.

Why do investors discount the inflationary portion of earnings growth? Because it's something that applies in more or less equal measure to all companies, not something special to that one company. The following analogy may help explain it. Suppose you have a group of young boys and you want to find the one who in a few years will be most likely to be a good basketball player. One of the boys is five inches taller than the others and appears to be continuing to grow at a faster rate. You're impressed. Now suppose that the school all the boys attend raises the floor of its gym by two inches every year. So all the boys, including the tallest, are in effect gaining these extra inches. But you wouldn't pay much attention to this aspect of growth because it's affecting all the boys equally. You still are most interested in the one whose real growth is outstanding. It's real growth that matters.

There is another more general reason why P/Es fall during inflationary eras: the increased uncertainty attached to earnings estimates. When inflation is low, businesses can look ahead with confidence and reasonably expect to carry out their plans for growth. Investors, too, can put faith in those plans. In inflationary times, though, you never know what's coming next. Time horizons become shorter, and profits less certain. Everything can change at a moment's notice, which makes investors understandably more wary. The upshot: further pressure on P/Es.

The 1970s versus the 1990s

We hope you now understand why rising inflation drives down P/Es. But what does this mean for you as an investor? What does it tell you about what lies ahead and what you should do about it? There are two ways of approaching an answer to this question—logic and the historical record. They both lead you in the same direction.

The first thing it means is that we are headed for an era in which stocks in general will be struggling, fighting to keep their heads above water, because as inflation rises, growth rates that used to be good enough to make stocks rise will become increasingly inadequate. To give another school-based analogy, it's as if you have a teacher who keeps tightening his standards for grading tests. The same level of effort that last month got you an A now earns you only a B. To keep getting A's, you need to study harder and harder. If you just maintain a steady level of work, your grades will keep slipping. So it is in times of inflation: growth rates that used to push stocks higher now no longer make the grade as investors become stricter in assigning P/Es.

Look at what happened in the inflationary 1970s, the period that the years ahead are most likely to resemble. At the start of the decade inflation was around 5 percent; by the end it was in double digits. Corporate profits, with a boost from inflation, grew at an annualized rate of more than 10 percent a year, well above average, and for the decade had nearly tripled. But P/Es, responding to inflation, fell in half, from nearly 16 to below 8. And stocks, as measured by the S&P 500, gained less than 20 percent for the decade.

Moreover that 20 percent vastly overstates how well investors actually performed. The gains in the S&P 500 worked out to less than 2 percent a year—while inflation averaged above 7 percent a year. Stocks were a losing proposition. Even counting dividends, the returns from stocks trailed inflation by nearly 2 percentage points a year. Any way you look at it, despite much-above-average profit growth, stocks were extremely poor investments during the 1970s, and the culprit was the high and steadily rising inflation that led to declining P/Es.

In fact, as figure 10a, "Inflation and Markets," shows, inflation can wreak more havoc on stocks than even a major economic cataclysm like the Great Depression. The period between 1966—when inflation began to rise—and 1981—when inflation began to retreat—was the worst fifteen-year period ever in the stock market, and that includes the 1929 crash and its aftermath. In the fifteen-year period 1929–44, stocks, despite declining more than 70 percent from their peak, still managed to post an average annualized gain—including both price changes and dividends—of more than 1.5 percent, above the near-zero inflation rate that prevailed for most of that time. For the 1966–81 period, even taking dividends into account, stocks had an average annualized gain of 6 percent, one point under the average of inflation for the period. This means that in 1966–81, investors in the S&P 500 lost more than 1 percent a year in purchasing power, compared with a roughly 1.5 percent gain in purchasing power during 1929–44. (While this makes the point that an investor did

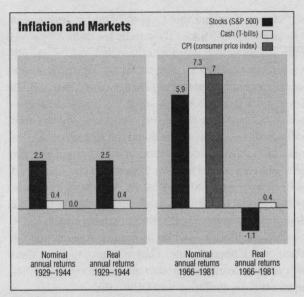

Figure 10a

worse during the inflationary 1966–81 period, it might seem like a fairly small differential. But over the fifteen years, the differences mount up. An investor who put $1,000 into the market in 1929 and stayed in for fifteen years emerged with about $1,250 in buying power. In 1966, the same strategy would have left an investor with less than $860 in purchasing power after fifteen years. That's a swing of 45 percent, a big deal indeed.)

In periods of low and falling inflation, everything operates in reverse. Let's look at the market between 1981 and 1999, a period when inflation fell from over 10 percent to about 2 percent. During those years, corporate profit growth averaged about 5 percent a year, less than in the inflationary 1970s and about in line with historical averages. Just how well did stocks do in that time? The S&P 500 rose nearly tenfold. The credit for this generation-long bull market goes to the lower inflation rate, which boosted P/Es from about 8 to over 30. This gain in P/Es meant that instead of rising two and a half times—their gain in profits—stocks chalked up tenfold gains.

In those low-inflation years, and particularly in the 1990s phase, rising P/Es were so much the engine of stock market gains that some of the biggest market winners had almost no earnings at all, just the ability to make giddy investors believe that great growth lay ahead. In the years ahead, though, stocks no longer will be able to make it on P/E appeal alone. They will have to prove their worth through earnings growth that is strong enough to counteract rising inflation and falling P/Es.

What to Buy: An Overview

Comparing the go-nowhere market of the 1970s with the sizzling market of 1981–99, we used a broad-based market index, the S&P 500, to make our point. With the S&P 500 representing about 80 percent of the value of all U.S. stocks, this was a convenient way to show the overall impact of inflation on stocks. Moreover, it makes it clear that in inflationary times you should avoid investing in any system that simply "buys the market."

Remember, the market is not a monolith. In the 1970s, while stocks overall suffered a small loss in real terms, individual groups bucked the prevailing inertia in a big way. For instance, the oil service companies gained 31 percent a year and gold rose 33 percent a year. Figure 10b, "Sailing Through the 1970s," gives a rundown of many of the groups that flourished in those inflationary years.

This time around, too, there will be wide divergences within an overall lackluster market, and some groups will surge powerfully. These winners will have one thing in common—unusually strong growth rates that will make up for the general downgrading of P/Es. Actually, we haven't pointed out in so many words, though it's implicit in what we've said, that in inflationary times earnings go up for almost all companies, because the inflation rate

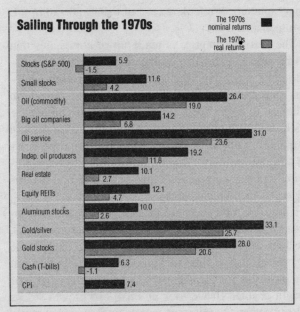

Sailing Through the 1970s

The 1970s nominal returns ■
The 1970s real returns ▨

	Nominal	Real
Stocks (S&P 500)	5.9	-1.5
Small stocks	11.6	4.2
Oil (commodity)	26.4	19.0
Big oil companies	14.2	6.8
Oil service	31.0	23.6
Indep. oil producers	19.2	11.8
Real estate	10.1	2.7
Equity REITs	12.1	4.7
Aluminum stocks	10.0	2.6
Gold/silver	33.1	25.7
Gold stocks	28.0	20.6
Cash (T-bills)	6.3	-1.1
CPI	7.4	

Figure 10b

gets tacked on to their growth rate. It's just that for most companies, these rising earnings aren't enough to compensate for falling P/Es. For the winners, though, earnings come on so strong that the drop in P/Es is overwhelmed.

What's their secret? In the years ahead, two broad and in some instances overlapping categories of stocks will be hard-wired to benefit from inflation and the energy crisis so as to generate the necessary stellar earnings growth. The first category is made up of companies whose earnings are leveraged to inflation in underlying commodities—such inflation beaters as oil and other energy companies and gold mines. Selecting stocks from these primary groups is the first key to outsmarting inflationary markets. We look at each of them in detail in following chapters.

The second category is made up of companies that will be able to capitalize on the coming energy crisis— that in one way or another will benefit from all the problems stemming from oil shortages and our long-term addiction to fossil fuels. In addition to oil companies, which straddle both categories, these also include defense companies, alternative energy companies, and property and casualty insurers. (One of the latter group is a multifaceted stock so compelling for so many different reasons that it gets its own chapter.)

These are the two broad categories of stocks that investors should focus most closely on in coming years. But there are other investments that are also likely to be very rewarding or that are important for rounding out your portfolio and providing a way for you to hedge your bets. In chapter 16 we look at small-cap growth stocks and real estate investment trusts, two groups directly leveraged to inflation. In addition, we spotlight a select handful of individual companies that are simply too attractive to pass by. Finally we spell out the worst of the worst—the investments that should be avoided at all costs. In chapter 17 we present deflation hedges, such as zero coupon bonds. These offer essential protection against the inevitable periods of recessionary jitters that arise during inflationary times.

Suspending Buy and Hold

Some investors follow their portfolio's ups and downs as avidly as a dog eyeing a piece of steak held in its trainer's hand. They have their broker's number on speed dial and are ready and willing to buy or sell at a moment's notice in an effort to maximize their gains. Other investors by temperament or training have absorbed the notion that you should buy and hold for the long term. When pressed, they point to studies that show that over time stocks always go up, plus they hate paying brokers' commissions. They are the ones who could take the proverbial trip around the world and never check on the status of their holdings.

Some of the greatest investors of all time have adhered in large measure to the buy-and-hold school. And in many market environments, it's an approach we endorse for most investors. Timing the market is always tricky and can easily backfire. There's nothing sadder than a sold-out bull watching a market take off somewhat ahead of when some timing system had suggested it would.

Having said that, we need to repeat our second rule of inflation investing—that in periods of high and rising prices, it pays for all investors to be willing at times to sell a hefty chunk of their holdings and switch into safer harbors. When inflation is dictating the economy's twists and turns, you still might ultimately come out ahead if you follow a strategy of buy and hold. But you lose so much ground and sacrifice so much of your potential gains that it's worth taking a more aggressive approach.

The reason is that in times of inflation, financial markets become far more volatile, subject to more frequent and sharper zigs and zags. And that's because, as we noted earlier, inflation generates tremendous amounts of uncertainty, for businesses and consumers alike. The future becomes murky and anxiety increases. It's harder to plan ahead when you don't know what your costs will be, what your inventories will be worth, and so on.

Further uncertainty comes from the fact that when inflation is running high, the threat hovers that the government may take action to bring it down, for example, by raising interest rates. No matter how carefully the government proceeds, it still might trigger too much of a slowdown, and in any case, fear of this happening can hurt the market too—leading to those periods of deflation or deflationary scares we've referred to. But if inflation has become embedded in the economy, the effects of any government action are likely to be only temporary. It's like putting a piece of cardboard on top of a saucepan containing a furiously boiling liquid. The cardboard may hold in the steam awhile, but unless you sharply turn down the flame, it won't do much good for long.

Once inflation passes 4 or 5 percent a year it tends to take on a life of its own. It becomes a vicious circle in which businesses raise prices to keep up with increased wages and the rising costs of raw materials, and those higher prices put further pressure on wages and raw materials. At any time, though, this trend can be temporarily interrupted by recessionary impulses.

In sum, once inflation becomes entrenched, the economy becomes inherently less stable. And this is reflected in far greater and more abrupt swings in the market.

In the high-inflation years from 1966 to 1982, stocks went through six—count 'em, six—bear markets, defined as stretches of time in which the major market averages declined 20 percent or more from their highs. By contrast, between 1982 and 1999, there were only two bear markets; moreover, each was brief and resulted in very little net change in the year in which it occurred. In fact, despite the market crash of 1987 the major averages still finished the year up. Only in 1990–91 did a bear result in a nominal loss in the major averages.

Here's the clincher for why in inflationary times buy and hold is a poor strategy: not only did the inflationary 1966–81 period experience a significant number of bear markets in the major averages, but during these times even the best, most inflation-leveraged investments got slammed. Oil service companies, for instance, had several bone-rattling drops of 20 percent or more. If you try to hang tough and hold on to your inflation-leveraged investments during such bear periods, you're simply giving away too big a portion of your potential gains.

In the following chapters we look in more detail at the particular groups and stocks that will enable you to sail through the turbulent years ahead in fine form, and we tell you exactly when and how to use our oil indicator to strengthen your competitive edge.

Key Points:

- ◆ Rising inflation drives P/Es down across the board.
- ◆ As inflation rises, most stocks will struggle, losing ground in real terms.
- ◆ In coming years investors should avoid the broad middle of stocks, such as indexes tied to the S&P 500. Instead, focus on groups leveraged to inflation and rising energy prices—primarily real assets.
- ◆ In the years ahead, shun a buy-and-hold approach. Use our oil indicator to switch between inflation and deflation plays.

CHAPTER 11

Energy

It may seem ironic and/or puzzling that our first category of recommendations is made up of today's best oil and natural gas companies. After all, we're predicting the eventual demise of oil and gas as our chief energy sources, so you might think these companies should be shunned. If fossil fuel production really is in the process of peaking, and reserves on the verge of declining, wouldn't oil companies—whose most important assets are fossil fuel reserves—be a bad bet? It would seem to be like deciding to buy tracts of forest just when the trees were being chopped down, never to be replaced.

It's a good question, but it has a good answer. Oil and gas companies, like any other kind of company, are in the business of trying to maximize the value of whatever assets they have. When oil prices are static or declining, the best way for them to do this is to produce all they can. But if oil prices are rising, the goal of maximizing profits would lead oil companies to husband their oil so as to get higher prices for it later. And this is exactly what we are likely to see. Oil and gas companies will try to produce just enough oil and gas so as to ensure that their revenues—fuel production multiplied by price—and their profits are in a long-term uptrend.

Some companies will be better at this than others. Those with the largest reserves will have an edge, but shrewd management also will be a critical ingredient in the smart handling of assets. The bottom line, though, is that well-positioned oil and gas companies—those with large reserves per share and good management—will be able to capitalize in a leveraged way on rising oil and gas prices.

Indeed, in the past several years many of the major oil and gas companies have already started to take this new tack. They have stopped trying to maximize production and instead are aiming for higher returns on capital. Return on capital is another way of saying profits, and in this case, it's another way of saying that the companies, rather than producing all they can, are saving some production for later when prices are higher.

One further point: in making the case for investing in the oils, an additional consideration is that some of today's big oil companies are ideally positioned to become leaders in alternative energies, the industry of the future, and already have gained stakes in this area.

The bottom line: oil and natural gas stocks should be core holdings in every investor's portfolio.

We first began recommending oil stocks in our last book, *Defying the Market,* which came out in 1999. Noting that oil prices had been in a prolonged slump and that oil stocks had long been the dogs of the market, we predicted that was about to change. In the years since then, as oil prices began their uptrend, oil stocks have solidly outperformed the market.

We're not citing our good call, and it was a very good call, just to brag. The point is that there were good reasons for investing in the oils then, and there are good reasons now. Their recent outperformance is just a prologue to what lies ahead.

Just how important is it to own energy stocks in inflationary times? If you go back to figure 10b on page 161, you'll see it shows annualized rates of gain for the oil industry and its various segments during the 1970s and compares this performance to that of the S&P 500. As you can see, the differential is huge. In the 1970s the S&P 500 had anemic average annual gains of under 6 percent a year, and in terms of real buying power actually lost ground. The majors went up 14.2 percent a year. The starkest comparison, though, is with the oil service companies, whose market gains averaged a staggering 31 percent a year. After inflation, this came to a still hefty 24 percent a year. With compounding, anytime you average 24 percent real gains a year over a ten-year period you end up multiplying your money more than eightfold. In other words, by the end of the decade you can buy more than eight times as many goods and services as you could at the beginning.

Moreover, these comparisons actually understate the difference. That's because the S&P 500 includes the major oil and oil service companies. If the comparison were made with the index minus its oil component, the contrast would be even greater. Without the oils, the S&P 500 gained less than 5 percent, not 6 percent, and lost even more ground to inflation—your purchasing power would have declined by more than 30 percent. Not much of a bargain. While the gains for the oil service sector were the highest, every part of the oil industry significantly outperformed the market at large.

What to Buy

Oil companies fall into three categories. Most prominent are the big diversified majors, household names such as Exxon—the companies that are thought of as big oil. Next are the oil service companies, a varied bunch. They include the drillers, which mostly build and operate rigs, as well as diversified service providers, which offer a range of exploration and drilling services to the majors. Finally, there are the independents, domestic companies that simply explore for and produce oil and natural gas.

Naturally, there are trade-offs. In the past, the oil service companies, in particular the drillers, were the riskier, more aggressive plays that offered the biggest potential returns. Under today's circumstances, though, as we'll discuss shortly, the oil service companies are facing more complex times. We'd concentrate largely on the two other sectors, treating the service companies as speculations. The large and diversified majors continue to offer the potential for strong inflation-beating gains in conjunction with considerable downside protection and steady dividend income. The independent or domestic oil companies probably offer even more upside potential, though it comes with greater risk.

The Majors . . .

The best-known major oil companies are Exxon, ChevronTexaco, Royal Dutch Petroleum, and British Petroleum. They are all huge by any standards, with operations around the world. For all of them, a major part of their business consists of what are known as upstream operations, that is, exploring for oil and producing it. But they also have so-called downstream operations, which means doing various things with the oil they produce. Refining, for example, is such an operation; oil companies take crude oil and refine it into gasoline. When oil prices fall, hurting profits from that end, their profits from refining tend to rise simply because one of their major costs has come down. Similarly, most of the majors make various chemical products. Because petroleum is a major component of these products, here, too, lower oil prices can boost profits. Downstream operations give them a lot of buoyancy to profit in both good times and bad.

In other words, the majors do well when oil prices rise and they don't do badly when oil prices fall. If you choose to invest in one or more of the majors, you'll participate in oil's rise and you'll also have a lot of downside protection for those periods when oil prices go down.

Another significant benefit is that the major oil companies offer excellent yields, typically more than twice that of the market. These yields are also exceptionally reliable. The only time a major oil company has cut its dividend in modern times was in the late 1980s when Texaco, which now is part of Chevron, was caught in a legal battle with Pennzoil and temporarily declared bankruptcy. Within two years, though, its dividend regained all the ground it had lost. If you want income that is both totally reliable and almost certain to rise, plus a chance to profit from rising energy prices, you definitely should invest in at least one of the majors. They offer it all in one easy package.

If we had to pick just one major oil company, ChevronTexaco, with operations in more than 180 countries, would be our first choice because of its relatively low P/E and its high dividend yield. In addition, it should continue to benefit as the 2001 merger of the two companies, Chevron and Texaco, becomes increasingly well integrated.

More speculative than the other majors but with more upside potential is a company that we didn't mention in the list above but that has all the pluses of those companies with an added currency kicker. It is PetroChina, the huge Chinese oil producer, whose ADRs (American Depository Receipts) trade on the New York Stock Exchange. We cover it in more detail in chapter 15, where we discuss the reasons behind Warren Buffett's decision to take a significant stake in the company.

In the years ahead, investors should also keep a careful eye on Russian oil companies. For now, however, in our view the political risks associated with Russia make investing in Russian companies too risky for all but the most seasoned pros.

. . . Oil Service Companies . . .

We noted above that we're less enthusiastic about the oil service companies than we were in the past. In past periods of rising oil prices and inflation, they were the industry sector most leveraged to rising oil prices, and they made the most spectacular gains, though intermittently they also would suffer stomach-churning drops. An inherently volatile group, they were the feast-or-famine choices. They encompass two distinct groups, the drillers and the diversified service companies. The drillers are the more narrowly focused companies, building rigs and leasing them out. They don't own a thing except their rigs, and depending on demand their revenues and profits can fluctuate wildly. The diversified oil service companies provide rigs and drilling services to the majors, but the bulk of their business lies in offering other services, such as seismic and geological analysis and management. Historically they have been more leveraged to gains in oil prices than the majors and less risky than the pure drillers.

The reason that for now we are somewhat more cautious on the oil service companies, and particularly the drillers, is that the argument we made above concerning the big oil and gas producers—that they will be able to maximize their revenues and profits by managing their reserves, waiting for higher prices—doesn't apply to the companies that drill and service wells. With oil companies no longer pulling out all the stops to increase production, the drillers and service companies could be less in demand than in the past. Moreover, there is no long-term scarcity of drilling rigs or drilling services. And even as oil and gas producers, by dint of maximizing their profits, end up with more money to spend, there is no guarantee they will spend it on drilling. Rather, they are equally likely to use it to raise dividends, repurchase shares, or undertake some other means of increasing shareholder value.

Still, we would by no means rule out investing in the service companies. As aggressive plays leveraged to rises in oil prices, they merit a small place in the energy portion of many portfolios. Altogether there are less than a dozen major players in the oil services arena. The clear standout in the field is Schlumberger, the most profitable and dominant diversified oil service company in the world. It has a technological edge in the area of well services that won't be overcome by any of its competitors, and as we try to wring out every last ounce of oil from our depleted fields, its profits could be in a sustained uptrend.

Two of the drillers also are worth considering as speculations, particularly after a decline in their share price. Nabors is the world's best-managed oil and gas driller. Since emerging from bankruptcy in the early 1990s, the company has gone from being a marginal player to assuming a role as the world's largest land driller. Any major drilling project in North America will likely engage Nabors as a key participant, and while, as we noted above, the big oil companies may be more selective in how and when they drill, they are not going to cease drilling entirely. Also well managed is Noble, which, with stakes in oil and gas drilling in many parts of the world, has managed to ride out some of the down periods in the highly cyclical drilling business far better than most of its competitors.

. . . The Independents

The third group consists of independent domestic oil and gas companies. They are attractive because to the extent that they manage reserves effectively they are pure plays on rising oil and gas prices. As you'll recall, similar dynamics to the ones that have pushed oil higher have sent natural gas into a raging bull market of its own. The outlook for natural gas parallels the one for oil—many years of rising demand and tightening supply. Well-situated natural gas companies will see rapid earnings growth, the kind that will enable them to transcend falling P/Es, and they can be a good choice for a diversified energy portfolio.

If you're going to invest in one of the independents, you can't do better than Devon Energy. Through a series of astute acquisitions and continued exploration success, this Oklahoma City–based company has emerged as one of the largest and most dynamic of the North American independent oil and gas producers. Since 1993, revenues have grown from under $100 million to more than $7 billion—an unmatched record in the energy patch.

Its exploration programs span the globe, from the Gulf of Mexico to China. With reserves divided almost equally between oil and gas, the company will benefit from price rises in either commodity. Exceptional technology, especially in horizontal drilling, has enabled it to increase its reserves significantly in fields viewed as showing little promise. The company also excels at managing its finances. With its strong cash flow it has managed to sharply cut its debt load, a legacy of past acquisitions.

Admittedly, Devon's size will make meaningful reserve additions—other than through acquisitions—tough to come by in the future. Even on this score, though, it is worth noting that the company has more than 400,000 unexplored acres that could result in a big payoff. But even without further large production increases, rising energy prices should translate into long-term growth of better than 10 percent.

No matter what measure you use—P/E, price-to-replacement value, or free cash flow yield—the stock is cheap relative to the market and relative to its peers. By virtue of its size and highly skilled management, Devon is one of the surest beneficiaries of rising energy prices.

Energy Funds

Rather than pick and choose among individual energy stocks, you might prefer to invest in a mutual fund that specializes in energy. There is absolutely nothing wrong with this; it's a question of temperament, time, and how much control you want to have over your investments. Our favorite energy fund is the Vanguard Energy Fund, a relatively safe and value-oriented pure play in the energy arena. Charging lower management fees than other energy funds, it has provided steady returns for more than a decade. Most of its holdings are in large integrated oil companies, but it achieves diversification by also investing in such subindustries as drilling, equipment and services, and gas pipelines.

For a less conservative fund, we'd suggest the ICON Energy Fund. Its strategy is to rotate its holdings among subsectors so as to take advantage of underpriced industries with potential to outperform within the overall energy field. As a result of this strategy the fund has a much smaller stake in the major integrated oils, which for other funds tend to smooth out the risks, and a more aggressive bet on smaller and riskier stocks. While the fund has a relatively short history, its performance has been excellent, and it could be a good way to complement investments in big oil.

One final pick is the Excelsior Energy and Natural Resources Fund, which focuses on non-energy natural resources along with energy. Its managers are willing to make sector bets—for instance, its swing toward gold-mining stocks in 2002 helped protect assets. Like the other funds we have presented, it has a relatively low expense ratio, no front or back loads, and fairly low turnover. All three are a convenient way to invest in energy relatively safely and inexpensively.

When to Get Out of Oil Too

We've mentioned before, but it's worth repeating, that paradoxically when oil prices rise sharply (a year-over-year change of 80 percent or more), triggering a negative signal from our oil indicator, it will pay to get out of a big chunk of your oil investments. You might think that sharp rises in oil prices would be manna for any oil company, raising profits even as the rest of the world suffers. But it doesn't work that way.

As we showed in chapter 1, sharp rises in oil prices can shake the economy to its core. The massive bear markets of 1973–74 and 2000–2002 began with a 100 percent increase in oil prices. Even oil stocks were hurt, and for good reason. Sharply rising oil prices threaten the entire economy—threaten to send it into a deflationary spiral in which demand for everything, including oil, plunges. In other words, investors view sharply rising oil prices as containing the seeds of their own destruction.

They realize that even the most strenuous government efforts may fail to keep economic growth going.

When the oil indicator flashes a negative signal, don't cling to oil. In fact, if you've chalked up big profits in your oil holdings, it's even more important that you protect them by selling a significant portion before you lose them all. When the portents turn favorable again, our oil indicator will get you back in plenty of time.

We have one final word of advice on how best to invest in oil, or rather on what not to do. Don't even think about investing in oil futures. Typically futures trading is highly risky, with low margins and short time frames, and it is not for the individual investor. We could be dead right on oil over the long term, but a relatively small downdraft in the short term—which could mean days or even hours—could wipe you out. Leave this particular form of investing to professional traders and managers of commodity companies. And this advice applies not just to oil but to gold and other commodities as well.

Key Points:

◆ Well-managed major oil and gas companies should be core holdings for all investors. They will parlay higher oil prices into higher return on capital. They also are positioned to become leaders in alternative energy.

◆ Oil companies were among the star performers of the inflationary 1970s.

◆ The safest are the majors. Most of the money you invest in energy should be divided among the big majors and the domestic producers, though aggressive investors can devote a small portion of their funds to the more speculative but potentially high-return oil service companies.

◆ When our oil indicator turns negative (deflationary), lighten up on your energy holdings as well, because sharply rising oil prices threaten the economy, including oil companies along with it.

CHAPTER **12**

Gold

Greedy old King Midas didn't yearn for dollars, yen, or euros. He wanted gold. He must have been anticipating inflation.

Gold is the quintessential inflation hedge. When prices are rising rapidly, and the value of paper money is slipping, people turn above all to gold. That has been true in the past, and it almost certainly will be true in the coming decade.

In fact, there is good reason to think that within a fairly short time, gold, which in the 1990s was one of the most dismal investments around, will be soaring. And we mean soaring. It's not far-fetched to think that within a few years gold may reach $1,000 an ounce. We even think it could reach $2,000. That's if we're right and inflation heats up for all the reasons we've been outlining—along with the emergence of some other economic conditions that we expect, like negative real interest rates. But even if inflation stays far tamer than we expect, gold is still poised to move considerably higher. Like the energy companies, gold—in the form of gold stocks—should be a core holding for all investors. Gold is almost certain to be a good investment, and it could be on the verge of a historic bull run. You won't want to miss out.

The Last Laugh

Some of the most colorful investors have always been the so-called gold bugs. Deep in their souls they believe that Armageddon is approaching and that the only defense is to hold gold and gold stocks. They congregate at various investment conferences, where they reinforce one another's fanatical devotion to the yellow metal, and they've been deterred not a whit by the inconvenient fact that for years they've been wrong. But pity them no longer. In the fairly near future, finally, their loyalty to gold will pay off, at least for those who have not passed on to a better place (gold bugs, as a rule, tend to be on the elderly side).

And in the world of investments, if you have a long-term perspective and are going to be loyal to anything, gold is as good a choice as any, and probably better than most, because over the long haul gold holds its value relative to inflation. That is, the same amount of gold that would buy an acre of property in ancient times would buy a comparable acre today. Gold has been prized as a source of value for some six thousand years and has been a formal currency for nearly three thousand years. And this is more than an arbitrary happenstance. The metal has some special properties that make it intrinsically beautiful and for practical purposes eternal. It doesn't tarnish, it is extremely malleable, and it is almost indestructible.

But if over multidecade and multicentury periods gold keeps pace with inflation, what makes it really interesting at this juncture is that it doesn't do so in a steady, evenly paced manner but rather with breathless lags and leaps. That is, when inflation is low and far from the minds of investors, gold doesn't just patiently mark time, it tends to sink in a big way. When inflation rises, particularly when some other key conditions are met, gold gallops ahead.

During the 1990s, gold was in one of its profound slumps when it dramatically underperformed inflation. Today, the signs are that all the conditions are falling or soon will fall into place for gold to let it start to make up lost ground, and then some, in a big way. The gold bugs who waited out the 1990s may well have the last laugh.

In fact, even during any deflationary interludes in coming years, gold is likely to outperform stocks, whose real returns would be dramatically negative. Gold, because it tends to track changes in general prices rather than changes in profits or other economic variables that influence stocks, would decline but would likely do so in line with the decline in overall prices.

A Quick Recap of Gold's Recent History

Into the mid-1970s, gold's price was fixed at $35 an ounce. It wasn't until 1975 that it began to trade freely. Gold opened 1975 trading at $200 an ounce. By the summer of 1976, it had dropped to $100 an ounce. No one much wanted gold; investors were far more interested in the stock market, where the Dow was flirting with 1000, a decade-long resistance level.

But the Dow once again failed to pierce the 1000 level, and in fact entered a mini bear market. And gold began to gain some traction, advancing at a moderate pace. Inflation was at 5 percent, though declining from the double-digit levels that had marked the early part of the decade. In 1977 inflation once again began to pick up steam, and by 1978 it reached 9 percent. Gold flew by its previous high of $200, and the metal and gold stocks began to sharply outperform the major market averages. They were killing bonds, which were mired in a major bear market of their own.

In 1979 inflation, whether measured by the consumer price index, the producer price index, or other gauges, hit 10 percent, and gold really took off. By the end of the year the metal had topped $500. Gold stocks, meanwhile, had nearly doubled from the start of the year and were nearly three times higher than they had been at their lows in 1976.

(Note that there is no automatic relationship between changes in the price of gold and changes in gold stocks. Sometimes, as in the 1970s, the gains are roughly parallel. But a lot of variables affect a gold stock's performance— the discovery of new reserves, for instance, or technological improvements that lower the costs of mining.)

In early 1980, in a wild speculative spasm, gold reached $800. Then it retreated, spending most of 1980 trading between $500 and $700. Gold stocks, though, continued to rise throughout the year. By the time they topped out they had risen nearly two and a half times from year-end 1979 levels and were about eight times above their 1976 lows.

That counts as one of the greatest bull markets of all time. Hard as it may be to believe, the bull market in tech stocks in the second part of the 1990s pales in comparison. From its 1996 low to its 2000 high, the Nasdaq, which is the best proxy for the tech stocks, grew less than fivefold. For the gains in the Nasdaq to have been comparable to the gains in gold, the Nasdaq would have had to have reached over 8000 instead of 5000.

In the 1980s, gold declined all the way back down to $400 an ounce, which is where it started the 1990s. In the 1990s it slid further, ending the decade under $300. Gold stocks did even worse, with the average gold stock losing nearly 50 percent in a decade in which the stock market as a whole was making extraordinary gains.

Inflation, Psychology, and Real Rates

It would be simple enough to say that we expect inflation to rise dramatically and that gold is an inflation hedge, and to let it go at that. But that leaves a lot of questions unanswered about gold's recent record, and it doesn't give us any basis for projecting just how high gold might go. Above we threw in the notion of $2,000 gold. That wasn't an arbitrary figure. To see why it is a very real possibility, we need to look in more detail at how gold and inflation interact.

In the 1990s, inflation averaged about 2.7 percent a year, just a bit below its longer-term average of about 3 percent. But gold and gold-related investments, as we noted above, didn't come even close to keeping up, with the metal sinking to $300 and gold stocks being cut in half.

Why were investors so disrespectful of gold during those years? True, inflation was at very moderate, not fear-mongering, levels. But the reasons go beyond that. In the 1990s investors were on a high in which they thought they had stumbled into the best of all possible worlds. They felt that real growth not only would remain strong for as far as the eye could see, but that it would accelerate. The reason they believed this was that they had bought into the myth that productivity, thanks to the presumed revolution in tech, not only was high but would continue to rise indefi-nitely. With productivity rising, inflation would continue to decline even as profits continued to rise.

If you believe all this, as investors did, you'd be nuts to buy gold, or anything other than financial assets such as stocks and to a lesser extent bonds. No wonder that by the end of the giddy 1990s all measures of market valua-tions, from P/Es to dividend yields, had shattered previ-ous records. And no wonder that relative to inflation, gold probably had one of its worst decades ever.

In other words, over relatively short periods of time— and from a historical perspective a decade is a short pe-riod of time—gold or any other investment can get totally out of whack with its underlying valuation benchmarks. That was emphatically the case for gold in the 1990s.

Now let's backtrack to the second half of the 1970s, in particular to 1979, when gold began rising in a big way. Inflation, as we noted, had reached 10 percent. But something else had occurred that was even more signifi-cant for gold than the actual level of inflation. In 1979 real interest rates turned negative. That means, as we've discussed before, that inflation was higher than interest rates, in this case higher than both long-term rates (bonds) and short-term rates such as T-bills. Negative real rates were the true trigger for gold's dramatic gains in 1979.

Remember, when real rates are negative, it means that you can borrow money, buy any asset that keeps up with inflation, and turn around and sell it for a real profit. Suppose, for instance, that inflation is 20 percent and interest rates are 10 percent. You buy a house for $100,000. A year later you sell it, and since its value has gone up in tandem with inflation, you get $120,000. All you have to pay the bank that holds the mortgage, though, is $110,000. You pocket a profit of $10,000. If that isn't coining money, we don't know what is.

Of course, you don't want to have to buy and sell a house every year, and there would be other costs involved such as taxes and maintenance. The trick is to find some asset that is reasonably liquid and that is sure to keep up with inflation. In addition, you have to be confident that while you hold this asset, real rates will remain negative. We'll get to the second condition in a moment, but as far as the first condition goes, is there any asset more likely to hold its value in times of inflation than gold? It is, as you probably realize, a rhetorical question. Negative real rates, even more than inflation alone, make gold a can't-lose investment with an appeal that feeds on itself.

The Not-to-Worry
Worst-Case Scenario

We're entering a period when the portents for gold are becoming as favorable as they get. Investors have stopped viewing the world through excessively roseate glasses. Even though productivity is still thought to be rising, no one thinks it is rising at ever faster rates. Inflation right now is still relatively low, but despite the emergence of some talk of deflation, by some measures prices are starting to rise. And hovering close by are all those inflationary pressures we have been describing—rising energy prices, accelerating government spending, and the burden of huge consumer debt.

Gold has one final factor on its side. It has been so undervalued that even if it just gets back to where it should be in order to be fairly valued it would mean a huge rise. In other words, the outlook ranges from excellent to extraordinary.

Our best assessment tends toward the extraordinary end of the range. We think that within fairly short order inflation will be rising and real interest rates will turn negative for long stretches of time as policymakers take care not to short-circuit economic growth. These conditions will push gold into a sensational new bull market. But before we explain our calculations for gold's best case, let's look at why gold is almost certain to rise sharply even if we're wrong about inflation, real rates, and all the rest.

For this minimalist case, all you need to agree to is that the 1990s were a decade of, to put it politely, unrealistic expectations. It's not a question of expecting doomsday but simply of acknowledging that no new paradigm has magically arisen and wiped out all the old economic relationships of the past. In other words, you just need to accept that in "normal" economies some inflation is likely to coexist with economic growth.

Once you agree to the above statements—and if you don't, then you're in utter denial about the hideous bear market that ravaged stocks in the first two years of the twenty-first century—a little math and a few straightforward assumptions make the case for gold. The most important assumption is that gold eventually ends up maintaining its value relative to inflation. We'll also assume that gold's price of $400 at the beginning of the 1990s was a relatively fair value. Gold at $400, after all, represented a more than 50 percent decline from gold's peak in 1980 and was a price that occurred after nearly a decade of declining inflation. In other words, there is no reason to think that gold was overvalued at the start of the 1990s relative to its long-term historical trends.

If $400 was a fair price in 1990, what should gold be in the early 2000s? Between 1990 and 2003, inflation averaged about 2.7 percent, which means prices have climbed more than 40 percent. Tack this on to $400, and you get $565. With gold in the low 300s in mid-2003, that's a pretty decent target. Moreover, it is a target that will rise right along with the inflation rate. That's not bad. But it's nothing compared to what gold is more likely to do.

The Aim-for-the-Fences
Best-Case Scenario

There are several reasons for believing that $565 is a very conservative target for gold. The first, of course, is that inflation is not likely to stay meekly in line with its 3 percent historical average. Rather, it is likely to accelerate sharply. And if and when it does, you can bet that gold will not just rise in line with rising prices—it will move well beyond inflation itself, as investors become as blackly certain that inflation will continue to rise as they were complacent in the 1990s.

Even more important is that real interest rates are likely to turn negative for much of the coming decade in order to help keep home prices from sagging and leading to economic collapse. And as our analysis, as well as the experience of the 1970s, demonstrates, negative real rates are what send gold into a frenzy.

But what about the question we posed earlier—whether investors will be able to count on real rates remaining negative for long enough periods of time? We think they can. Negative rates could yield to positive rates if policymakers act to bring inflation down, but the process won't happen all at once, nor would it have an instantaneous effect on gold. When Volcker raised interest rates in 1980, rates did turn positive and inflation came down. But as we noted above, gold continued to shoot up, not stopping until it hit $800. And even after it dropped, gold stocks remained stellar performers for most of the rest of the year.

Why did gold stocks continue to shine even after the metal itself had dropped by as much as $200? One reason is that rather than reacting to a trend in a commodity's price, stocks often react to absolute levels of a commodity, which are what count as far as earnings go. While gold was down sharply from its high, it still was well above the levels of 1979.

In any case, this time around we expect real rates to remain negative for all but brief periods of time for many years to come. This means the economic backdrop for gold over the next decade or longer could be at least as positive as it was for gold during the 1970s and for tech stocks during the 1990s. Remember, the key thing to watch is not so much inflation but the level of real interest rates. Gold began a bull market early this decade with inflation still at very low levels but with real rates close to zero and headed below zero. Of course, the higher inflation goes, the more room for very high negative real rates.

Keep another thing in mind—in 1980 the driving forces behind inflation were largely temporary conditions that had readily available solutions. Though energy prices were surging, new sources were coming onstream in the form of the North Sea and a large number of nuclear plants. Also, the Fed was facing a far less leveraged world and thus had more latitude to raise rates.

With all this as background, we can plug in some numbers. And if we're right on oil, inflation, and the need to keep real rates mostly negative, gold's rise will be stunning. To see why, let's look at the 1980 peak in more detail. In 1980 there were about 220,000 tons of gold in the world, of which a little more than half had been mined. The value of all this gold—below and above ground—was then about $5 trillion. Gold's durability—remember, it is almost impossible to destroy the metal—makes it possible to compare the value of the world's gold across different time frames. By April 2003 the value of all this gold had declined to about $2 trillion.

Now let's view these numbers in relation to financial assets. In 1980 the market capitalization of all stocks in the S&P 500 was about $1 trillion. In April 2003 the value of all S&P 500 stocks was about $8 trillion.

One way to conceptualize these figures is to say that when gold fever was at its peak in 1980, gold was valued at five times the S&P 500. If we applied this ratio today, it suggests that gold will have a twenty-fold gain—which would bring it to more than $6,000 an ounce.

But this is somewhat simplistic, because almost everyone would accept that the underlying value of the companies that constitute the S&P 500 is greater today than it was in 1980. It's as if you were valuing gold in terms of all the diamonds in the world and then found out that a lot more diamonds had been discovered in the meantime—obviously, you wouldn't expect the same ratios to apply.

So how do we get a handle on the increase in the underlying value of stocks since 1980? One rough-and-ready approach would be to factor out the influence of inflation. Since 1980 prices as measured by the CPI have climbed about 2.5 times, while stocks rose eightfold. Divide 2.5 into 8 and you get a little over 3. That constitutes the real gain in the underlying value of stocks—a triple.

Note that you get roughly the same figure by using 5 percent as the average annual rate of real profit growth for the S&P 500 between 1980 and 2003. That implies a slightly more than threefold gain in profits and hence in the real value of stocks.

So however you calculate it, it's reasonable to figure the underlying value of stocks has increased threefold. This suggests that you should take our $6,000 figure above and divide it by 3. Presto: you get $2,000 as a reasonable target for gold, assuming inflation heats up as we expect and real rates are generally negative.

And if this sounds wild, remember once again that while in 1980 inflation stemmed from relatively ephemeral factors, in today's world the causes are deep-rooted; inflation may last longer and go higher, and real rates may stay negative longer. A target of $2,000 is wholly consistent with these realities.

Investing in Gold

The record from the 1970s suggests that gold stocks and the metal perform more or less in line with each other. Because stocks are a lot easier to buy and sell than, say, gold coins, we recommend that you participate in gold's move through individual gold stocks or gold mutual funds.

Barrick Gold, headquartered in Toronto and the world's second-largest mining company, is one of the greatest success stories in the history of gold mining. From a low of about $.50 a share in the mid-1980s, the stock climbed to more than $30 a share in the mid-1990s before retreating. During that ten-year period gold itself was relatively stagnant, and the typical gold stock turned in a lackluster performance. Barrick was the exception because it was able to find additional gold deposits. While those days, unfortunately, are behind it, Barrick still is a well-managed company whose stock price should faithfully follow the price of gold in the years ahead. It is an excellent alternative to buying the metal outright.

Even bigger than Barrick is Newmont Mining, the Denver-based gold-mining company that should be a core position in any gold portfolio. The company is known for not hedging its gold production, which means its earnings are more volatile than those of other gold mines but also more leveraged to changes in the price of gold. Because the company is so big, it is unlikely that Newmont will discover new gold deposits that are sufficiently large that they will add in a meaningful way to the company's value. Still, like Barrick, the stock should rise in line with the price of gold in years ahead, and this should be reward enough for those who invest in the company.

When it comes to investing in mutual funds specializing in gold stocks, one problem is that as gold fever began running high in the early 2000s, gold funds attracted so many new investors that many closed their doors to new money. The investors who bought in time were lucky: in 2002, gold-oriented funds were top performers, posting total returns that averaged 63.28 percent. We can't promise that any of the funds described below will be open to new investors or that they will rack up similar gains in the future, but they are definitely worth knowing about.

The Tocqueville Gold Fund focuses on selecting undervalued gold mines that don't hedge. This strategy propelled the fund to the head of the group as gold prices were rising; when gold prices fall, however, the fund is likely to lag the group. The fund also owns more small-cap and micro-cap companies than most of its counterparts, which further adds to volatility.

Two other and very similar funds are the Gabelli Gold

Fund and the American Century Global Gold Fund. Both are pure gold funds and both are invested in a mix of large and moderately sized mines located primarily in North America and South Africa. Gabelli Gold also owns some gold bullion.

One of the cheapest and most conservatively managed funds is the Vanguard Precious Metals Fund. By investing mostly in larger, lower-cost gold mines as well as in producers of precious metals other than gold, the fund has kept volatility relatively low.

A Word About Hedging

You may read at various times that a particular mine, including any of our recommendations, is hedging some of its production, and some analysts may suggest this is a negative thing. But don't get too alarmed.

When a gold mine hedges, it agrees it will sell some of its future production at a price somewhat higher than whatever price exists at the time the contract is entered into. Hedging assures a company a fixed cash flow. The downside is that if in the meantime gold rises to levels higher than the contract price, the mine can't capitalize on that rise.

But the calculus is a little more complex than the above suggests. For instance, the gold company might use the cash it gets to develop new properties that in the long run might more than make up for the differential in price on the hedged gold. A company also might hedge if it thinks gold prices will sink during the time it takes to produce the gold that is being hedged. The point is that good managements use hedging when they think it is appropriate and likely to increase returns over the long term. The key is to have faith in management, and the companies we have picked have managements that have proved adept over considerable periods of time.

Key Points:

- ◆ Gold, the quintessential inflation hedge, could reach $1,000 to $2,000 an ounce or higher within the next few years.
- ◆ Over the very long term, gold has kept pace with inflation. Shorter-term, though, it can lag behind or leap ahead. In 1976–80 gold stocks soared eightfold, a bull market that dwarfs the 1990s' rise in tech.
- ◆ Even more than inflation, negative real interest rates—which we expect in the years ahead—are the impetus for big gains in gold.
- ◆ Gold stocks and gold funds are the easiest ways to participate in gold's uptrend.

CHAPTER 13

Alternative Energy

Oil companies and gold investments will benefit from the problems we are facing—rising energy prices and accelerating inflation. Our next group of recommendations will benefit from efforts to find solutions to the problems. They are companies that in one way or another are players in the alternative energy/conservation/supplemental energy arena. It's an area that we expect will expand and blossom over the next decade, and thus in many ways our recommendations here can best be viewed as an initial stab at getting a toehold in a young and promising field.

Our picks range from a giant and diversified blue-chip company whose stake in alternative energies is just a very small part of its overall operations to a small, speculative company built entirely around one innovative process. We also include under the rubric of alternative energies several silver and other metal companies that supply materials vital to alternative energy technologies. One point is that there are a lot of ways to go about investing in alternative energies, involving lots of different levels of risk. And a second point is that if you look at companies with the idea of alternative energies bobbing around in the back of your head, you may find that the alternative energy angle shows up in some surprising places.

In recommending some of these alternative energy companies, we are, in a way, taking a leap of faith. We're saying that oil will grow scarcer, oil prices will rise, and, ergo, alternative energies will come out of the woodwork and into the limelight. Companies will want to develop them, consumers will want to use them, and government will agree to fund them. Is this realistic, or are we taking too much for granted? As investors, is it prudent to bet on the group just because a concerted and rapid push to develop and switch to alternative energies makes eminent sense?

We can't say for sure. Actually, we can almost say for sure that at some point this sector will come into its own, we just can't say exactly how it will happen—through what exact combination of government policies and private investment—or that it will happen right away. We still think, however, that most investors should include some alternative energy stocks among their holdings. If the path that will lead to the ascendancy of alternative energies isn't crystal clear, the big-picture reality is that alternative energies will be directly leveraged to the central economic reality of the next decade and longer: an energy crisis leading to a long-term rise in oil prices.

Moreover, while genuine alternative energy plays— those based on true renewable fuels—may be somewhat more of a gamble, two of our recommendations are in what we have termed the supplemental energy area, that is, more conventional oil alternatives. These require less of a leap of faith—in fact, they are probably as sure a thing as any investments around.

Below we discuss a diverse group of stocks with some play in the alternative energy/conservation/supplemental energy area. One nice thing is that most of them also meet at least some of our basic criteria for investing in inflationary times. You can think of them as alternative energy plays or as energy and inflation plays with an alternative energy kicker. The one stock that is an exception to our rules has overwhelming compensating strengths.

A Low-Risk Bet on Wind

We'll start with the exception to our inflation-investing rules—General Electric. It's not a real asset company, it isn't directly leveraged to inflation, and with a revenue base of $130 billion, it's definitely not a small-cap growth play. It's also not generally thought of as an alternative energy company. But nonetheless it's a terrific choice for most portfolios and offers a low-risk and potentially highly rewarding way to take a position in one of the most compelling alternative energies for the near and intermediate term.

That's because in 2002, for $225 million, the company acquired Enron Wind, putting it in the forefront of what down the road is likely to become a burgeoning and massive market for wind energy. Enron Wind is a vertically integrated producer of wind energy: it generates wind energy on its own wind farms and is a leading manufacturer of wind turbines and blades. While currently its revenues and profits would barely constitute a rounding error on its parent company's balance sheet, it is likely to be making major contributions to General Electric's bottom line within a short time. And within the decade it could be the driving force in the company's overall growth, if wind energy, as we expect, takes off and becomes a multi-hundred-billion-dollar industry.

As we detailed in our chapter on alternative energies, when it comes to wind energy, the technology already exists, and wind can generate electricity more cheaply than either coal or natural gas. This cost advantage virtually ensures strong growth in this market over the long haul. What's keeping wind from making immediate massive inroads into the electricity market is the enormous investment in infrastructure required. If government were to assume a major role, as it did in creating the interstate highway system, growth in wind energy would explode. What does "major" mean? An investment in the hundreds of billions in today's dollars would be comparable to what the federal government put into highways. This should be a minimum commitment.

Earlier we offered our thoughts on why wind should become a major force, no pun intended, in our energy mix in the near future. While we're not counting on our views to carry the necessary weight when it comes to determining policies, we do think that the facts will eventually speak for themselves and that inevitably at some point the government in conjunction with private industry will make a large-scale commitment to wind energy. We hope it is sooner rather than later. The thing to realize is that electricity sales in the U.S. currently top $220 billion. Within the next ten years wind energy could be a hundred-billion-dollar industry. And General Electric will be a clear leader.

There are, of course, other reasons to like General Electric. One of the world's best-managed companies, it is a leader in a multitude of industrial and financial markets ranging from turbines to jet engines to leasing. Its long-term growth record is unmatched among major industrial companies. One statistic that stands out is its record of more than twenty-five consecutive years of dividend increases. Over the past decade dividend growth has been well over 10 percent a year.

It is true that if alternative energies do not drive GE's growth to the high levels we expect, the stock could suffer the same fate as other high-P/E stocks and end up a mediocre performer. But we are willing to bet that with its exceptional capital base, the company will indeed succeed in leveraging its growth to our necessity for new energies. Given its major place in our economy, betting on General Electric is akin to betting on the future of capitalism in the U.S. In today's world it becomes an especially compelling bet when you throw in the company's leading stake in a key to capitalism's survival, alternative energies. For a low-risk, income-producing stake in alternative energies, you can't do better than General Electric.

Four Additional Mainstream Plays

For all the safety and contamination issues that ultimately will limit nuclear energy's use, nuclear power is likely to become more prominent in the relatively near future. The reason: it will gain a major and growing cost advantage as natural gas, the current marginal source of electricity, becomes less readily available and more expensive. For this reason one of our picks in the supplemental energy area is Exelon, which, headquartered in Chicago, is the largest producer of nuclear power in the U.S. Because design and construction costs are things of the past for Exelon, it can produce electricity far more cheaply than utilities relying on other fuels. Based just on its existing operations, dividend and earnings growth should be around 10 percent a year. Moreover, it's possible that despite the huge cost of building new nuclear plants—stemming in large part from the need to address environmental and safety concerns—at some point as fossil fuel prices rise it may start to make economic sense to construct new nuclear facilities. Exelon would be the most likely manager of any new nuclear plants. This would clearly boost profits even more. We consider Exelon an attractive choice for all investors.

Two Canadian energy producers are double plays: they qualify both as energy stocks and, because of their position in tar sands, as alternatives to current fossil fuels. EnCana, based in Calgary, is the largest independent natural gas producer in North America, while Petro-Canada is an integrated Canadian oil company. In terms of their primary energy businesses alone, both are very attractive companies. Each is growing reserves and production much faster than its competitors, and each can reasonably expect to increase production by 10 percent a year over the next three to five years. Thus each will benefit not only from rising energy prices but also from rising production. This combination could translate into long-term growth of 15 percent or more a year using very conservative estimates about how high energy prices are going.

But the reason we've put them in this chapter is because of their stake in Canada's enormous tar sands reserves, which by some estimates contain as much oil as Saudi Arabia. We haven't mentioned tar sands up to now because we really don't expect them to play a significant role in solving our energy problems. Converting tar sands into usable oil causes a lot of pollution and requires tremendous amounts of water. Further, production costs are sensitive to the underlying costs of energy; they rise as energy prices rise. Even under the most optimistic assumptions, Canada's tar sands will be producing less than 2 million barrels of oil a day by the end of the decade. And that projection doesn't factor in any possible restraints that might arise from Canada's adherence to the Kyoto agreement's environmental standards. But if tar sands will be only a niche alternative to oil and gas, they nonetheless have the potential to make a big contribution to the bottom lines of the few companies that can produce and market them. For EnCana and Petro-Canada, their stake in tar sands is likely to be the added kicker that will make them standouts among energy producers.

Earlier, noting that alternative energies and conservation are generally thought of as two facets of the same solution, we argued that they really are quite different and that efforts to mandate conservation could prove counterproductive. But naturally occurring conservation efforts are a different matter. More than 90 percent of all the oil consumed in the U.S. goes for transportation, with cars gobbling up the lion's share. Thus, by definition, significant conservation will have to start with cars. And as gasoline prices rise—whether strictly because of higher oil prices or from higher taxes as well—drivers increasingly will crave more fuel-efficient cars.

Somewhat ironically, then, from an investment angle the best conservation play is an automobile company, not the first thing that springs to mind when envisioning alternative energies. Our choice is Toyota. Japanese automakers already have been luring American car buyers with such highly fuel-efficient cars as Toyota's hybrid Prius. But Toyota is a standout not just for its proven ability to adjust to changing consumer demands but, maybe even more, for its extraordinary financial position. The company's balance sheet is a marvel: not only is it by a wide margin better than that of any other carmaker, it is one of the best among all companies anywhere—one of only a dozen or so companies that Standard & Poor's rates as AAA. (General Electric, by the way, is another.) This massive financial advantage in tandem with its dexterous management will give Toyota an enormous edge as consumers start to demand more fuel-efficient cars. Moreover, its offerings of such cars will give the company a major boost in the emerging market for cars in countries such as China and India. Toyota is likely to be a brilliant standout in a pallid automotive industry.

A Rank Speculation

The smallest, most speculative alternative energy stock we are recommending is Utah-based Headwaters. It trades on the Nasdaq and has a market capitalization of below $500 million. Normally, in fact, in a book—as opposed to a regularly published investment newsletter that can monitor its recommendations on a continual basis—

we wouldn't even mention such a relatively small and obscure company. The main reason we are doing so now is that Headwaters is perhaps the only company—and this is a sad comment on the state of alternative energy technologies—that is at once profitable and engaged in just the kind of innovative technology that will be critical in helping us to deal with our energy problems as they become ever more intense.

Headwaters' major businesses revolve around converting coal into clean alternative fuels and making use of coal by-products. In 2002 the company entered into a multiyear, $2 billion agreement with the Shenhua Group, which is China's largest coal company. Shenhua will use Headwaters' proprietary technology to convert coal into ultra-clean hydrocarbons, which can then be refined into gasoline and jet fuel. The goal is to eventually produce 50,000 barrels of clean-burning gasoline a day. (It is worth noting that if we extrapolate from these numbers, we can infer that a $200 billion investment in coal conversion in this country could result in 5 million barrels of gasoline a day, reducing U.S. oil imports by nearly 50 percent. This would go hand in glove with investing in wind energy, as we proposed in chapter 7. Again, it is a question of money and the will to use it, not technology.)

The ability to use coal as a relatively clean substitute for oil clearly could be a major thrust of efforts to deal with declining oil supplies and rising oil prices. The agreement with China could take several years to pay off, and there is no guarantee as to the degree to which the company will share in the spoils. Still, Headwaters has a lead in what may be a critical technology. If it maintains its lead—and with small companies that is always a tremendous if—the payoff could be enormous. If this stock intrigues you, our advice would be to follow the stock—any broker can keep you updated—and if the contract with Shenhua remains on track, take a small position, knowing it is a speculation. If Headwaters does falter, though, still keep your eyes peeled for any other company that has a viable coal-to-gasoline technology.

The Silver Lining

Our three final alternative energy recommendations are also precious metals plays, giving them the double-whammy appeal that we really like. One is a silver producer, and the other two are producers of what are called platinum group metals, or PGMs, a group that includes platinum, palladium, and related rare metals.

All these metals have unique properties and a variety of uses. Silver and platinum, of course, like gold have long been prized for their beauty and durability. Silver is also the best conductor of electricity, making it a critical metal in many electronic applications. In addition, it has medicinal uses and is widely used in photography. But in terms of alternative energies, the most exciting technology for silver revolves around the transmission and storage of electricity through high-temperature superconducting (HTS) wires.

A little background here is useful. Almost a generation ago two scientists, Johannes Georg Bednorz and Karl Alex Müller, won a Nobel Prize for their discovery that certain materials could conduct electricity without any resistance at temperatures well above absolute zero. Before that, superconductivity had been achieved only at temperatures close to absolute zero. To keep wires that cold required the use of prohibitively expensive liquid helium. HTS, by contrast, can be achieved by using liquid hydrogen, a far less expensive substance. Still, most of the hopes associated with HTS have not been brought to fruition.

One dream, however, still seems to hold real promise: using HTS wires to transmit and store electricity. The significance is that HTS wire eliminates all friction, which means that none of the electricity that travels along it is wasted as heat. The potential savings in energy are substantial. As natural gas prices continue to go up, this technology should become more cost-effective. Additionally, because electricity actually can be stored in HTS wire for later use, it may prove valuable in enhancing the use of alternative energies where intermittency can be a problem, for example, with wind.

Silver comes into play because for HTS wire to be supple enough to function properly, it must be coated with silver, quite a lot of it. One expert in this field, Paul Bateman of the Silver Institute, has estimated that HTS wire has the potential to consume more than 50 million ounces of silver a year, equal to about 20 percent of current industrial demand.

Even a lot less than that would significantly increase demand for silver. And as natural gas prices rise, 50 million ounces could prove a conservative goal. Silver inventories have been declining—by 2003 they were at less than half their 1992 levels. That's because the demand for silver has been met from existing supplies, since at current prices it hasn't paid for silver mines to produce more of the metal. Yet as supplies drop, silver prices, depicted in figure 13a,

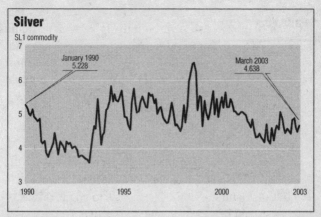

Silver

SL1 commodity

January 1990
5.228

March 2003
4.638

Figure 13a

"Silver," have continued to stagnate because investors have been expecting demand to fall. The awareness of a new and major source of demand is likely to reverse this perception and send silver prices flying.

There is ample precedent for a long period of sideways trading in a metal followed by an explosive multi-year breakout. As figure 13b, "Palladium," shows, that metal is a prime example. For years palladium hovered near $150 an ounce, and then within a short period of time shot up to the $1,000 area as protracted and growing demand for the metal triggered an abrupt rise in price. Such sudden breakouts are examples of "catastrophe" theory at work. This is the notion, developed by the late René Thom, that a major change may occur seemingly all at once and with no prior warning in response to one small final cause. Or, to invoke a more popular expression of this theory, it's the idea of the final straw and

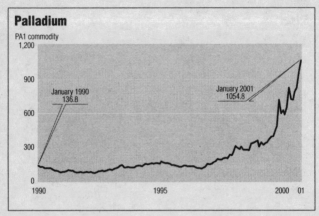

Figure 13b

the poor camel's back. For silver, the straw could well be the recognition of the metal's essential role in energy conservation.

If you invest in silver, you'll be in good company. In 1997, Warren Buffett reported that he had accumulated a major portion of the world's silver inventories. His announcement created such a stir that he subsequently stopped reporting his silver positions, but if the extent of his current holdings is subject to speculation, there is no doubt at all as to why he bought the metal. He stated that it was because of a chronic imbalance between supply and demand. That imbalance has widened dramatically since 1997.

What is the best way to invest in silver? Unlike gold, there are only a few silver stocks and no silver funds. One approach is to buy the metal directly. To do this, go to the

Web site www.kitco.com. However, we suggest that you allocate no more than half your total silver investment to the metal itself. For the rest we recommend investing in our favorite silver stock, Apex.

Apex, which is traded on the American Stock Exchange, stands out both for its massive silver holdings and its potential to add to those holdings. The performance of a metal stock relates to two things: the price of the metal and how much of it the company has on hand. Usually reserves don't move enough to affect the stock price. But there are exceptions. In the 1980s, for instance, Barrick Gold, the gold mine we recommended last chapter, was an outstanding performer even though the price of gold was declining sharply. The reason: its discovery of huge amounts of the metal. Apex could be a similar story, making it a highly leveraged play on rising silver prices. Its major holding is its silver mine in Bolivia, which is one of the world's largest, with total reserves estimated at more than 450 million ounces of silver.

We have other reasons for having confidence in Apex, even though, as a small-cap mining stock, it is a higher-risk choice than we are generally recommending. One is the quality of the company's chairman, Thomas Kaplan, a Ph.D. from Oxford whom we know personally. What impresses us about Dr. Kaplan is that he recognizes that silver is a scarce commodity and with his historian's perspective refuses to mine it until prices break out. And second is the quality of some of the investors in the company, who include George Soros as well as Moore Capital, one of the largest and most successful hedge funds.

One final point: investors like Buffett and Soros don't

make bets unless they are expecting something big. A growing recognition of silver's role in helping us cope with a worsening energy crisis could be the spark that sends this critical metal up as much as tenfold. Apex would be a prime beneficiary.

The PGM Syndrome

Platinum group metals are rarer than silver and at least as critical to the functioning of the world's economies. They are irreplaceable catalysts in many chemical processes. Probably their best-known use is in catalytic converters, the devices in cars that help clean up noxious emissions. Automakers have been trying hard to find less expensive substitutes for platinum or palladium for this purpose, but nothing else has come close. Thus until we replace the internal combustion engine with new nonpolluting technologies, PGMs will be in demand for their environmental protection capabilities.

But not to worry—when the internal combustion engine goes the way of the dinosaur and the kerosene lamp, PGMs will still be in demand. That's because they are also essential ingredients of hydrogen fuel cells, which are the likely candidates to someday replace internal combustion engines in the cars of the future.

Of course, as we noted earlier, fuel cell technology has a way to go. Fuel cells are still very expensive. Moreover, while the idea is that they don't pollute because they simply combine hydrogen and oxygen to create electricity, with water the only by-product, as we've discussed earlier, right now the only way to isolate that hydrogen is to burn fossil fuels. Still, if or when economies of scale bring down the costs, fuel cells will begin to make inroads in the automobile industry. And this will be a boon for PGMs, which are the only practical catalyst discovered so far capable of bringing about the chemical reaction that combines oxygen with hydrogen.

Once PGMs become more widely associated with a potential solution to the energy crisis, their value will increase, if only for reasons of speculation. As we noted earlier, hydrogen fuel cells have been gaining attention. Typically investors interested in this area focus on companies like Canada-based Ballards, a developer of fuel cells. But the risk with a company like Ballards is that ten years is a long time and who knows if its technology will carry the day. We'd go another route. The thing that is known is that if fuel cells become more widely used, the demand for PGMs—whose value is likely to increase anyway in line with rising inflation—will get an extra boost. So if you want to bet on fuel cells, we'd suggest investing in PGM mines or buying platinum coins.

One problem with investing in PGM mines is that the major supplies are in less-than-secure parts of the world. By far the biggest producers are in the southern African nations of South Africa and Zimbabwe, as well as the former Soviet Union. The few North American plays are small, struggling companies. Our recommendation for a stock would be South Africa–based Impala Platinum. Because of its location there are political risks, but it has a relatively high yield and is substantially leveraged to PGM prices.

If you prefer to buy platinum coins, we'd recommend doing so through the same Web site mentioned in conjunction with gold and silver—www.kitco.com.

There is one other potential application of PGMs, in this case, specifically palladium, and while we know it is very remote, we think we might as well mention it. And that is its use if cold fusion ever got off the ground. We haven't even discussed cold fusion, which seemed to hit a dead end in the early 1980s. But research is continuing, and while it's a long way off from producing definite results, we wouldn't rule out something eventually coming from it. In March 1999, a director of the prestigious SRI International research center presented the results of experiments about which he said: "It may not be nuclear fusion. But a new, clean source of power may, in fact, be on the horizon." The point is that all efforts so far to bring about cold fusion have involved palladium or related metals. As we said, we thought it was worth mentioning because, as the ad says, hey, you never know.

Conservation Misplays

The investments recommended above will benefit as rising oil prices spur interest in both alternative energies and conservation. Earlier we explained why we hope the main emphasis will be on alternative energies, arguing that efforts to impose conservation are likely to be self-defeating. But obviously some people would totally disagree.

Most proponents of conservation focus on its perceived ability to help us cut oil usage and lessen pollution and environmental damage. The brilliant Amory Lovins, a scientist and environmentalist, however, has taken it a step further. He has passionately argued that conservation not only would cut our use of fossil fuels, it would raise standards of living and enable corporations to raise profit margins. If you are aware of his writings, or learn of them, you might wonder if it wouldn't make sense to invest in companies that aggressively practice conservation—not to reward their efforts at social responsibility on moral grounds but out of a selfish desire to reap some hefty rewards yourself.

It's an appealing thought, but unfortunately there is no evidence that such companies are likely to benefit from their commendable efforts to practice conservation through recycling and other means. The chief example cited by Lovins in his most recent book, *Natural Capitalism,* was a relatively small industrial carpet manufacturer called Interface, which at the time was riding high. Lovins claimed that Interface, through its emphasis on saving and reusing material, had garnered a tremendous edge over more wasteful competitors. The history of the company since then, though, makes this seem unlikely. Since 1997, when Interface was being hailed as the bright new face of capitalism, it went from being very profitable to struggling to survive. The stock price fell by more than 65 percent, and the company lost considerable ground to its rivals. We don't think that such companies are the way to go. To capitalize on the growing awareness that oil's days are numbered, stick with the types of investments we've outlined in this chapter.

Key Points:

◆ At some point the alternative energy/conservation/ supplemental energy area, which is leveraged to one of the key realities of the coming era—scarcer oil and rising prices—will take off in a significant way.

◆ Alternative energy stocks come in all shapes and sizes. They range from blue-chip giant General Electric, with its big stake in wind energy, to a small speculation like Headwaters.

◆ Other picks that will capitalize on diminishing oil and the need for alternative energies: Toyota, silver companies, and platinum group metals.

CHAPTER 14

Defense

Alternative energies are our hope for the future. But it's clear that no matter how frantically we try to develop and switch over to them, we won't do so overnight. Meanwhile, our economy still needs energy, and lots of it, and for now that means it still needs oil. This fact was the major reason we recommended that oil stocks make up a core part of every investor's portfolio. And it's also the reason we are recommending defense stocks.

For those of you who thought or hoped that the end of the Cold War meant defense expenditures in the U.S. could remain on a permanent downtrend, the reality is that we need a strong defense now as much as ever. Only the nature of the threat facing us has changed. When our enemy was the Communist world, a strong defense was needed to protect us from being physically dominated. Today, a strong defense is all that keeps us from economic catastrophe. It's all that ensures that we will have a strong call on the world's increasingly insufficient supplies of oil.

Unstable Producers

If that sounds either far-fetched or melodramatic, it's not. As worldwide oil supplies grow scarcer, and as large areas of the world, such as China, continue to industrialize, meaning they become ever more avid consumers of energy, countries will be competing with one another to buy the oil their economies require. This is a big change, and countries that have no means of throwing their weight around will lose out. For the U.S., military might is the ace in the hole that will ensure that we have access to diminishing supplies of oil—that our allocations receive favorable treatment from oil producers.

Wait a minute, you might be thinking. Don't we need a strong defense because of other things—terrorism, rogue nuclear states, and so on? Isn't that enough of a reason without dragging in a potential energy crisis?

Actually, no. Obviously, end of Cold War or no, the U.S. never had a thought of dismantling its entire military arm. Whoever is in the Oval Office, it never was and isn't likely ever to be a question of beating our swords into plowshares. But it's a matter of degree. If all we had to worry about were essentially political crises, we could maintain and upgrade our military capability at a relatively moderate level and still have the power and flexibility we need. It's the looming energy crisis that suggests we will need to build up our defenses at an accelerated pace. Defense stocks are the obvious beneficiaries.

To understand why energy and defense are so closely linked, consider the fact that almost all the oil in the world is produced in economically underdeveloped and hence intrinsically unstable countries. We're not talking only about Middle Eastern oil producers, though they obviously fit the bill. We're also thinking of such countries as Venezuela and Nigeria. In other words, our economy's lifeblood depends on an assortment of volatile countries with relatively immature economies.

Both these concepts—volatility and economic immaturity—are important. In the years ahead, every one of these countries is going to be essential to worldwide oil supplies. Civil unrest or a coup or anything that jeopardizes oil production in even one country is likely to have an impact that resounds throughout the oil-consuming world. For instance, the 2 million barrels of oil that Nigeria exports daily easily could be the difference between continued economic growth and a major recession. The U.S. will need the ability to intervene militarily, or to plausibly threaten to do so, in the event of any major disruption in any oil-producing country.

But that's just part of it. The other part has to do with the fact that because most oil-producing nations are underdeveloped, they currently are happy to export most of the oil they produce—they don't need it to run their own economies. To the extent that their economies become more developed, they will need to use more oil themselves, and the amount of oil they are willing to export will shrink. It's not inconceivable that at some point we'll be implicitly relying on overwhelming military might to ensure that they continue to export what we need.

Regardless of how quickly the economies of oil

exporters themselves become more industrialized, one indisputable reality is that the economies of several other huge underdeveloped nations have been moving quickly along such a path, complicating the task the U.S. faces of ensuring it will have adequate oil in the years ahead. China and India, for instance, both will be competing with us, and the competition will get increasingly fierce.

If you think we're being paranoid here, we're not alone in this assessment. Philip Abelson is editor emeritus of *Science* magazine, one of the world's most prestigious journals. Its lead editorial never engages in political haranguing and usually deals with issues of scientific research and funding. In the April 4, 2003, issue, however, Abelson wrote as follows about oil: "An ominous development has been the great increase in net imports from East Asia, including China, Japan, Taiwan, and Korea. These countries have been increasingly competitive in both high- and low-technology items. . . . In 1986 China . . . exported 500,000 barrels of oil/day. In 2002 the Chinese competed with the United States for imports of oil. It imported 1,300,000 barrels of oil/day. The Chinese economy continues to grow rapidly. China is emphasizing the training and retaining of far more engineers than the United States does. In the future, will China or the United States be able to obtain and pay for their imported oil?"

Of course, it would make sense to react to these realities by developing alternative energies as quickly as possible. But the quote is also a clear reminder that the only edge the U.S. possesses in the competition for oil is its overwhelming defense superiority. Imagine for a moment that it is China that has the world's greatest military might, not the U.S. Think how insecure we would be about the future of our economy then. And that insecurity would be justified, because without a strong defense there is little likelihood that in the years ahead we would get a decent share of whatever oil still is being produced around the world. It's not that we're likely to have to undertake military action against China or other potential big consumers of oil or against the oil producers themselves. But an indisputably strong military will be the quietly effective big stick, the absence of which would deflate our national psyche and place our economic prospects in dire jeopardy.

Dollar Signs

Oil today is priced everywhere in dollars, and in a world where the dollar can quickly drop or inflation quickly rise, this is a tremendous boon for the U.S. For example, between the summer of 2001 and the spring of 2002, the dollar fell 20 percent against a basket of the world's other major currencies. If oil exporters had been pricing their oil in terms of those currencies rather than the dollar, the U.S. would have been paying 20 percent more for its oil just like that.

There's another benefit to the U.S. from having oil priced in dollars: it helps maintain the dollar as the world's reserve currency, which loosely means that banks and corporations need the dollar more than other currencies to pay for internationally based transactions. If some other currency replaced the dollar as the world's reserve currency, the U.S. would lose a lot of control over its economic fate. Oil sales are probably the most important of all internationally based transactions, and the willingness of oil exporters to price oil in terms of dollars is a big reason that the dollar remains the world's reserve currency. We would argue that the chief reason oil exporters have remained so loyal to the dollar is their undoubted awareness of our military resources—again, the silently effective big stick.

And it's not inconceivable that oil producers might prefer to free themselves from pricing their treasured resource in dollars. In particular, suppose that everything we've predicted about rising oil prices and inflation comes to pass. Then all currencies in effect will be depreciating, some faster than others. In this inflationary world, commodity exporters might find it preferable to price their products in terms of gold—in the case of oil, say ten barrels of oil for one ounce of gold. If commodities were priced in terms of gold, the dollar, along with all other currencies, would lose some luster, and in effect rising oil prices could end up being even more inflationary to the U.S. We may be years away from such a possibility, but it is hardly implausible. Again, the reality is that even though the U.S. is the world's most important economy, it would be quite powerless to prevent all this from happening without a strong military.

Real Rises in Defense Spending

Defense companies are sure to benefit in a big way from all the above. Spending on defense, which slowed down in the 1990s—the so-called peace dividend effect—has been rising in the current decade and is likely to continue rising for quite a while longer. Of great significance, it will be rising not just in nominal terms but also in real terms. In other words, it will be rising faster than inflation.

Figure 14a, "Defense Expenditures," shows defense spending on new programs as a percentage of gross domestic product from 1978 to 2003. When the ratio is rising, it means that after inflation, gains in defense spending on research and for new programs exceed gains in economic growth. In 2000, such defense expenditures accounted for less than 1 percent of GDP. If, as we think likely, the figure rises to around 1.5 percent over the next decade or so, it would mean that in real terms defense expenditures grew 50 percent faster than the economy. That is an entirely plausible scenario—1.5 percent would take us to just half the level of defense spending at the height of the Cold War and would be below the ratio for most of the postwar period. Indeed, if you believe that a strong defense is as vital today as it ever has been—and we think that because of the energy/defense connection it is—you could argue this is a conservative target.

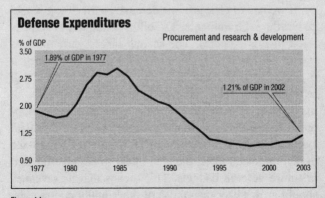

Defense Expenditures

% of GDP — Procurement and research & development

1.89% of GDP in 1977

1.21% of GDP in 2002

Figure 14a

When defense expenditures rise as a percentage of GDP, it is a multiple win for defense companies. In chapter 10 we detailed why P/Es for most stocks would be falling. But if a company's real profits are growing faster than inflation, and growing in a way that seems believable and sustainable, then that company can buck the trend. Its P/E can rise even as all around it P/Es are falling. Defense companies are likely to be one of the few groups that manage this hat trick. They are positioned to benefit from a potent combination of rising P/Es and rising earnings.

This has happened in the past. Between 1977 and 1980, as inflation rose and P/Es for the stock market as a whole were falling, defense expenditures were rising as a percentage of GDP, and P/Es for defense companies rose, in some cases sharply. Defense companies were among the market's leading groups. We think history will repeat itself.

The Best Defense

If we had to pick just one defense contractor, it would be Northrop Grumman, a $25 billion company headquartered in Los Angeles. The second-largest defense contractor and the largest shipbuilder for the U.S. military, it is poised to grow faster than any of its competitors well into the 2000s. The company is involved in virtually all major defense activities, ranging from missile development, to space research and systems, to mission-critical systems. Its 2002 acquisition of TRW gave it a huge stake in defense electronics, the area likely to be most favored in future defense budgets. This means that its revenues should grow faster than those of other major contractors.

In addition, synergies associated with the merger should boost profit margins. Rapid growth in cash flow and earnings will lead to the rapid repayment of the debt from that and other acquisitions, and as the decade matures will translate into big dividend increases or share repurchases.

More than half the business of Raytheon comes from electronics, which should stand the company in excellent stead as defense expenditures increase. Cash flow is high and almost certain to increase in the years ahead. There are some negatives, though. The company has a history of missing forecasts and of occasionally stumbling. In addition, its balance sheet is more leveraged than those of its competitors, and the company has unresolved issues related to the bankruptcy of two power plants once under its control. Still, investing in this company is probably the surest way to capitalize on the growing focus on electronics within defense.

One area of defense spending that will get a lot of attention in coming years is information technology. A company that should benefit from this trend is CACI, which has been doing government information technology work for more than forty years and derives more than 60 percent of its revenues from the Defense Department. Its profits should continue to grow considerably faster than defense expenditures. Its strong position results from the high-level security clearance it has earned and its strong relationship with the Defense Department. Another major positive is its nearly debt-free balance sheet. Its growth also will be spurred by acquisitions, which can be funded by the company's ample free cash. In fact, its free cash flow yield is likely to rise sharply in the years ahead. Over the past decade, even with the slowdown in defense spending, CACI still managed profit growth of 20 percent a year. It should offer comparable results for a while to come.

General Dynamics generally gets a bad rap because of its commercial jet business. But while this division certainly has been a drag on profits recently, it is just a relatively small part of the company, whose defense business is thriving. Indeed, the company's profits have grown by better than 15 percent a year since the late 1990s. Investors also have tended to mark down General Dynamics because its defense business doesn't focus on the glamour areas such as communications and technology. But its strong shipbuilding and marine systems are valuable franchises that should continue to generate double-digit growth in profits. Moreover, if the economy strengthens as we expect, the commercial airline business could turn out to be a leveraged positive for the company. The stock offers an excellent combination of growth and income.

Lockheed Martin, formed in 1995 from the merger of Lockheed and Martin Marietta and headquartered in Bethesda, Maryland, is the nation's largest defense contractor and has a leading or at least a large stake in every aspect of the defense industry. In 2002, nearly 80 percent of its business was with U.S. government agencies. While historically it has traded at a higher P/E than its rivals, it still is an excellent choice for most investors, ensuring them a broad-based stake in the defense arena. Its balance sheet is admirable, and like the other major contractors, the company generates a lot of free cash, which can be used to pay dividends, make acquisitions, or buy back shares.

Boeing is the world's largest aerospace and defense company, the nation's largest exporter, and one of the biggest spenders on research and development. It doesn't qualify as a pure defense play, as more than 40 percent of its revenues come from building commercial airliners. Because of the financial woes of the airline industry, this part of its business will be slow-growing and erratic, and rising energy prices could make the situation even worse. Still, Boeing has a major presence in many defense programs and could be a good choice for long-term investors.

Key Points:

◆ Defense spending is sure to rise in a big way to ensure our access to diminishing oil supplies, which are concentrated in underdeveloped, inherently unstable parts of the world.

◆ Defense expenditures will likely rise faster than inflation. This will be a multiple win for defense companies: not only will their profits rise, but they may also be among the few groups to experience rising P/Es.

CHAPTER 15

Berkshire Hathaway

And now for something completely different: an entire chapter devoted not to a broad group or category but to one stock. Not just any stock, of course—it's a giant that in many ways is like a group in its own right. It's a stock that should benefit in unique ways from our current dependence on fossil fuels and from the growing shortages that we foresee. Moreover, it's a stock that is at once both an inflation hedge and a deflation hedge and that offers a rare combination of frisky growth and bedrock safety. The stock, no surprise, is Berkshire Hathaway, the creation of one of the greatest investors of all time, Warren Buffett.

Berkshire Hathaway has long been a favorite of ours; we featured it in our last book, and, to say the least, it hasn't disappointed. It has been a stellar performer and we expect it will continue to be one for a long time. We think it should be a core holding for investors of all stripes. Let's look at how it should be able to parlay its strengths in the turbulent years that lie ahead and why we're giving it such raves.

Insurers in the Spotlight

You may have noticed that in a book focusing on the urgent need to move beyond fossil fuels, we've barely mentioned the environment. This might seem like a major oversight, given that a good deal of the impetus behind efforts to curb the use of fossil fuels comes from a desire to cut carbon emissions, which both damage health and almost surely contribute to global warming. The reason: we're convinced that it won't be environmental concerns that will lead us away from fossil fuels, but rather rising oil prices.

That said, the evidence is strong that global warming exists and may be causing even more damage than is commonly understood. In a May 16, 2003, article in *Science* that looks at flood risks in central Europe, authors Alfred Becker and Uwe Grunewald note that a warmer earth surface resulting from global warming will increase evaporation, while a warmer atmosphere can hold more water vapor. They predict: "The intensity of rainfall and the frequency of extreme events should, therefore, increase. Such an increase is one of the most critical threats of global climate change in the 21st century."

Whether you accept this or not, there is one thing we do know for sure, which relates directly to Berkshire Hathaway's prospects. Weather-related property damage around the world has been in a powerful and widely overlooked uptrend, in all likelihood the result of global warming. Figure 15a, "Weather-Related Woes," depicts

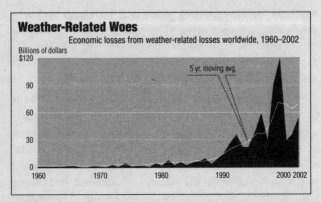

Weather-Related Woes

Economic losses from weather-related losses worldwide, 1960–2002

Figure 15a

worldwide economic losses caused by various outbreaks of extreme weather, from drought to floods to tornadoes, and the numbers are staggering. During the first five years of the 1960s, weather-related damage averaged about half a billion dollars a year. During the five years ending in 2002, the figure was close to $70 billion. For the entire period from 1960 through 1989, weather-related damage totaled $76 billion, which is less than single-year totals for the years 1998 and 1999.

Clearly, if this trend continues, it could cause widespread economic hardship and disruption that would hit at the earnings of a lot of companies and the pocketbooks of a lot of consumers. For instance, it could cause insurance rates to go way up, raising costs and adding to inflation; it could necessitate huge increases in government spending on disaster relief; it could force many companies out of business, raising unemployment; and so on. The destruction of the World Trade Center, while obviously a different kind of disaster, with special ramifications, hints at some of the economic dislocations that can result from any type of sudden catastrophic blow.

But if these dislocations will hurt large sections of the economy, they will help one group—the best-situated insurers.

Berkshire's Bona Fides: Insurance . . .

One company in particular is uniquely positioned to benefit from any and all weather-caused mayhem—Berkshire Hathaway, through its role as the best-capitalized property and casualty insurer in the world. Thus, as an investment, Berkshire Hathaway offers tremendous financial protection against the economic repercussions of any major weather-related disasters that may lie ahead.

Berkshire Hathaway got into the property and casualty insurance business in 1998 through its merger with General Re, the world's third-largest reinsurer, and reinsurance forms the bulk of its insurance business. Reinsurance is bought by insurance companies as a backup for their own capital in the event of a major catastrophe. Thus a reinsurer's strongest selling point is its capital base—obviously there's not much point in a company's trying to protect its own capital by buying insurance from a company that might have trouble paying off. Two factors are critical to the success of any reinsurer: its balance sheet and the skills of its management. On both counts Berkshire Hathaway stands miles above the pack.

For starters, Berkshire Hathaway is the only AAA-rated reinsurer in the world. But this high credit rating barely begins to tell the story. Berkshire's capital base of over $60 billion is more than 10 percent of the capital base of the entire insurance industry and more than 50 percent greater than the combined capital of its closest three competitors. This extraordinary capital advantage ensures Berkshire an unassailable position for a long time to come.

Here's why. When catastrophes occur, capital is drained from the industry as companies pay off claims. Some insurers go broke in the process. Others pull in their horns, writing fewer policies as capital is reduced. In 2002, insured disasters drained about 25 percent of the capital out of the industry, and one of Berkshire's closest competitors, Swiss Re, cut its dividend for the first time in nearly a century. Poorly performing financial markets have also been siphoning off capital from the industry as a whole as their investments have lost value.

Any event that takes capital from the industry will hurt the majority of companies but will help those that have the most capital, because they gain a competitive advantage. As less well-capitalized companies fall by the wayside, the survivors hold all the cards and can raise rates more than enough to compensate for the losses in capital that they themselves have suffered. While Berkshire has taken a few hits during the past few years, these were barely noticed because they represented such a minute portion of its massive capital base. In 2002, Berkshire's profits soared to a record $2,795 a share ($93.16 per class B share), more than five times the 2001 figures.

When it comes to management skills, Berkshire is also tops. Probably the best measure of an insurer's management skills is how much a company pays for its "float," which is the premiums an insurer receives and invests. The cost of the float is the difference between the premiums and all expenses, including insurance payouts. Many insurance companies frequently lose money on their basic insurance operations but make it up through their investments.

For Berkshire, however, in many years, and especially those in which financial markets were strong, such as the years between 1993 and 1998, its cost of float has been less than zero. (When financial markets are strong, all insurers have more money, and Berkshire gets far pickier about whom to insure, taking on only the least risky or most desirable customers. Thus its payout expenses tend to shrink.) In effect the company was being paid to use other people's money to earn money in strong financial markets. This is akin to coining money, and only the most skillfully managed insurers manage to do this on a fairly consistent basis.

It really doesn't get better than this. As an insurance company, Berkshire has more of everything than any of its competitors, so no matter what is happening, it is always the leader of the pack. In bad times—periods rocked by physical disasters and financial turbulence—Berkshire, because of its exceptional management and unmatched financial resources, makes money, big money, on insurance. When times are good, Berkshire makes money—big money—on its investments.

This is as close as you can get to a perfect hedge, a hedge that translates into strong growth under virtually every market and economic condition. And this includes both deflation and inflation. If the economy were to turn deflationary, Berkshire would stand strong because of its exceptional portfolio of bonds. But Berkshire's ability to make money when the stock and bond markets turn down is to some extent an inflation hedge, because, as you'll recall, inflation is a broad negative for financial markets.

. . . Deep Franchises . . .

If Berkshire Hathaway were nothing more than its reinsurance business, it still would be an overwhelmingly attractive company. But as you no doubt know, there is a lot more. In fact, it was a great company long before its merger with General Re, and while this merger fundamentally changed Berkshire's nature from a holding company to an operating company, it still maintains all the pluses from its previous incarnation.

Until that merger in 1998, you'd have to say that Berkshire's main asset was the genius of its head, Warren Buffett, who, disdainful of investing trends and conventional wisdom, demonstrated a preternatural ability to pick stocks and businesses that blossomed into major winners. Buffett bought Berkshire Hathaway in 1965 and turned it into a vehicle for buying shares in public companies as well as for gaining ownership of a number of privately owned businesses. Since then, the company's book value has grown from $19 a share to more than $41,000. That works out to a remarkable annual compounded gain of 22.2 percent, compared to 10 percent for the S&P 500.

Prior to 1998 you more or less could have duplicated Buffett's performance simply by buying the stocks that Berkshire was accumulating. Today his stock holdings are merely icing on the cake, secondary to his insurance operations. Still, it's worth looking at what this legendary investor is doing in the market today. With one important exception—PetroChina, which we discuss below—his stocks are the kind that can grow in almost all environments. They feature household names—Coca-Cola, American Express, Gillette, Moody's, the Washington Post Company, and Wells Fargo. One and all, they are deep franchises, which means that by virtue of their strong name identification they have tremendous control over their markets and over pricing within their markets. This control gives them an edge over the common run of companies during inflationary periods. They satisfy timeless common-sense criteria that have proved their worth again and again.

These are all great companies, and their records prove it. In 2003, for instance, their average growth was more than 50 percent higher than that of the S&P 500. And as we indicated, deep franchises like these by their nature are the quintessential all-weather investments. Their control over pricing means that when inflation is high they are more protected than most because they can raise prices with less resistance. And when inflation is low, their unit growth will stand out.

Still, we are not recommending them as individual investments. The reason is that their growth doesn't come cheap: in 2003 the average P/E was well above that of the S&P 500. In short, they are a bit expensive compared to the market, and as we have explained, this is a market that is going to be tough on high-P/E stocks. Even companies with superior control over their pricing will face increased uncertainty as inflation rises, and this will likely translate into falling P/Es. Buffett's picks are great companies, but we think you're a lot safer owning them through purchasing Berkshire, with all its other overwhelming strengths. However, if their P/Es drop to where they match the valuation of the market as a whole, they would be attractive investments in their own right. The key is not to overpay.

. . . And the PetroChina Gambit

In the spring of 2003, Buffett did something he had never done before: he took a significant stake in a pure commodity play, buying over half a billion dollars' worth of shares in PetroChina, China's largest oil and natural gas producer. In doing so, he appeared to be girding Berkshire in several ways to flourish in the inflationary times that lie ahead.

First, PetroChina will benefit from rising energy prices. But there's more: it also should benefit from other inflationary forces, including the rising U.S. deficit. That's because rising deficits and inflation in the U.S. will likely put a great deal of downward pressure on the U.S. dollar. The upward pressure on China's currency, the yuan, which as of this writing was pegged by the Chinese government to the dollar, is likely to be great. If China allows its currency to float against the dollar it almost surely will float higher, and this would benefit shareholders of PetroChina as it would automatically translate into a higher price for its shares.

PetroChina, therefore, represents an excellent investment that is likely to benefit from the most important trends we see developing over the next decade and more. Keep in mind, though, that it assumes China remains a stable member of the world community. Because of the potential for volatility in China, it should represent only a small portion of your overall portfolio.

One final point about PetroChina and Buffett's other stock holdings: the fact that overall they are not cheap doesn't lessen our enthusiasm for Berkshire Hathaway. Stocks represent much less than half of Buffett's overall capital, with the rest of it in bonds and cash vehicles such as T-bills that are largely immune to changes in the stock market. Even if all his stocks went to zero he would still have a massive edge in total capital compared to competitors in the insurance industry. The performance of his stocks is almost irrelevant.

In fact, we can go one step further. If the stock market fell sharply, while Berkshire would suffer some losses, its capital base would be far more intact than those of his competitors in the insurance business. It's another measure of how well this company is put together that a bad market could actually put it in an even stronger position.

Summing Up

Okay, we'll stop being coy. We like Berkshire Hathaway. In fact, it offers so much that if we had to pick just one investment for long-term gains in the years ahead, Berkshire would be it. It's truly an all-weather company, positioned to see its underlying value increase in times of either inflation or deflation. Over the long haul a stock's price almost always keeps up with value. Thus a long-term uptrend in Berkshire seems nearly assured.

Over the shorter term, too, whether one year or five years, you name it, Berkshire also would be our top pick. That's in large part because of its tremendous earnings momentum combined with its extraordinary balance sheet. In uncertain times, if you can combine outsized growth with safety, you're in an unbeatable position.

Of course, we don't have to pick just one stock, which is why we still think you should own energy companies, gold, and all the other things we've been urging. These choices are more directly leveraged to inflation than Berkshire and are thus more aggressive plays on the likeliest economic scenario. But by adding Berkshire, with its well-rounded assortment of unique strengths, to the mix, your portfolio gains in depth and balance.

What is the stock of Berkshire worth? We think a lot more than it was trading for in early 2003. Over the long haul virtually every metric associated with Berkshire, ranging from book value to share price, has outpaced the S&P 500 by a factor of two or more. Even more remarkable is the consistency of this performance. Only once between 1965 and 2003 did Berkshire's book value decline. During the total of five years in which stocks suffered double-digit losses, including the terrible bear markets of the early 1970s and more recently the early 2000s, Berkshire's book value increased in all but one year and had a total compounded gain during those five miserable years of more than 35 percent. This compares to losses in the S&P 500 of 72 percent.

Clearly, Berkshire is worth a tremendous premium to the market both for its exceptional growth prospects and its record of protecting investors when times are bad. Yet in terms of most metrics, such as earnings, book value, and cash flow, the stock trades at a discount or at no premium.

One quick note: as you probably know, Berkshire has two classes of stocks, A shares and B shares. The A shares are priced thirty times higher and investors in them have voting rights, but there is no difference in terms of core value. So it's your choice. As long as Warren Buffett, or a trusted handpicked successor, is at the helm, we wouldn't worry about the voting rights aspect. And if you are worried about what might happen to Berkshire if Buffett is no longer running the show, don't. The company is no longer dependent on stock-picking skills—though on that score, Buffett's underlings have demonstrated adeptness nearly equal to his. Rather, because of its massive capital advantage, Berkshire is assured, Buffett or no Buffett, of dominating the reinsurance business.

Berkshire Hathaway is a remarkable company that every investor should own. It is both an inflation and deflation play and thus the perfect all-around choice for the years ahead.

Key Points:

◆ Berkshire Hathaway is an all-purpose, all-weather investment that belongs in every portfolio.

◆ With Berkshire's unmatched capital base, its insurance business will flourish as environmental disasters continue to rise.

◆ At the same time, its strong portfolio of bonds makes it an excellent deflation hedge.

◆ Berkshire continues to hold a sterling group of dominant companies, such as Coca-Cola, American Express, and the Washington Post Company, while its big investment in PetroChina gives it a direct stake in rising oil prices.

CHAPTER 16

The Best of the Rest

We've given you the best of our inflation-beating and energy-shortage investment ideas. The groups and stocks presented in the preceding chapters should form the core of your portfolio in the years ahead.

Still, in addition to these categories, there are other inflation-friendly investments that you should know about and that may be particularly well suited to many portfolios. Moreover, while energy shortages and rising inflation will form the central axis of the economy in coming years, there will be coexisting trends that will spawn some irresistibly attractive investments, and there will be additional companies that simply stand out for one reason or another. It would be foolish to ignore these opportunities just so as to rigidly adhere to a particular investment agenda. Investors need to be pragmatists, not theologians. In this chapter we present this miscellany of worthwhile companies.

And finally, we take a moment to point out the worst

of the worst—the investments that, no matter what anyone may tell you, have no merit at all. After all, someone is bound to own them. Don't let it be you.

Small-Cap Stocks

Two categories of inflation-beating investments didn't seem quite major or compelling enough to warrant chapters of their own, but they still are good to know about and can have an important place in many portfolios. They are small-cap growth stocks and real estate investment trusts (REITs).

Small-cap companies generally are defined as those capitalized at under $1 billion. One index of small-cap companies is the S&P 600 Small Cap Index, a group of six hundred stocks whose average market capitalization at the start of 2003 was about half a billion dollars. Another index is the Value Line Index, which consists of more than sixteen hundred companies. While this index includes big companies as well as small, and indeed contains the great majority of big-cap stocks, it nonetheless is a good proxy for small-cap performance because it treats changes in all companies as equally important. That is, it will give a move in a small tech stock the same weight as a comparable move in General Electric. It's a true democracy.

In times of high and rising inflation, small-cap companies generally have an edge over the big caps. That's because, starting from a smaller revenue and earnings base, they can more easily maintain the rapid growth rates needed to overcome the drag of falling P/Es. If you are starting with just $1,000 a year in revenues, it's a lot

easier to keep doubling your growth than if you're starting with $1 billion and need to generate another $1 billion. Moreover, small-cap stocks are less risky in times of
inflation than when inflation is low. The ability to freely
raise prices, which is one characteristic of an inflationary
environment, can sometimes turn things around for a
small-cap company. In the inflation of the 1970s, small-
cap stocks dramatically outperformed big-cap stocks.

This could happen again. But you should also be
aware of the drawbacks that have kept us from explicitly
recommending small-cap stocks this time around. One is
their volatility. Fears of recession, never mind actual recession, can send them crashing. During the recession
scare of 1998, for instance, while big-cap stocks fell by
about 20 percent, the small caps shed more than 35 percent. Meanwhile, a real recession can lead to mayhem in
the small caps. During the 1973–74 bear market, while
big caps fell 45 percent from their highs, small caps gave
up nearly 80 percent.

Still, when inflation rules, small-cap stocks are likely
to be strong performers, and if you have a large enough
portfolio and a willingness to take a few extra risks, you
should consider them. The reason we mentioned the size
of your portfolio is that when you get into small-cap
stocks, you need to diversify heavily. Not only should
they form just a relatively small portion of your overall
holdings, you should have no less than ten different small
caps and preferably closer to twenty. This is practical
only if your portfolio is reasonably sizable.

By following certain guidelines you can cut the risks
of small-cap stock investing. The idea is to go for quality
and to ignore companies that have a good story but lack

any proof that they can live up to their claims. Above all, we would look for stocks with a solid record of earnings growth—if possible, earnings that have risen for at least ten years, or for as long as the stock has been publicly traded. Also important, look for solid balance sheets, with debt that is 20 percent or less of total capital. This gives the company some leeway if times suddenly grow lean. The company should have some free cash flow, which is money left over after all expenses are paid, as well as operating margins close to all-time highs, which signal that the company is dominant within a profitable sector of its industry. Finally, relatively low P/Es are a big plus. Specifically, we'd stick to companies whose annual earnings growth for the past five years has been no more than 50 percent lower than their P/Es.

Unless you're really into this kind of thing, doing all this research on ten or more companies may be a bigger job than you'd like to take on. An alternative approach would be to invest in a small-cap fund. Our first choice would be the Royce Total Return Fund, one of eleven small-cap funds under the management of the well-respected Royce fund family. Since its inception in 1994, the Royce Total Return Fund has consistently outperformed the small-cap benchmark index, the Russell 2000 Value. Generally holding 200 to 275 individual companies and with average volatility less than that of the S&P 500, this is probably one of the safest small-cap funds around. It invests in the securities of financially strong, dividend-paying companies trading significantly below their apparent value.

A second, more aggressive fund is the Third Avenue Small-Cap Value Fund, which consists of a smaller, more concentrated group of stocks. Although relatively young, the fund belongs to a fund family with long experience in the small-cap arena. It focuses on finding value among small-cap companies but will continue to hold or even increase positions in companies that grow into bigger-cap status.

The third standout among small-cap funds, the Fidelity Low-Priced Stock Fund, is technically not a small-cap fund at all, though it is generally followed as such. Rather, it invests in companies with a share price of $35 or less, but since most of these are small-cap or even micro-cap companies, the fund is a good way to gain a stake in the small-cap world. It has consistently been a top performer and offers broad diversification. Its charges are lower than those of most of its competitors.

REITs

If the coming period will be marked by the desperate need to keep home prices buoyant, it follows that real estate should be a consistently strong sector of the economy, and indeed, that is likely to be the case. The problem, of course, is that unless you're in the business, real estate tends to be a rather unwieldy sort of investment. Sure, you could buy a house on spec with the idea of reselling it in the near future for a nice profit, but that approach simply isn't practical for most people. And while it's nice to know that your own home may be appreciating in value, you probably would like to hold on to it because of such minor considerations as needing a place to live.

There are, however, ways to participate in a real estate boom without doing anything more demanding than calling your broker (your stockbroker, that is, not your real estate broker). That's by buying shares in a real estate investment trust, or REIT, which you can do as easily as you'd buy shares in any publicly traded company.

The great majority of REITs are known as equity REITs, and these are the ones we would stick with, because they will benefit directly from strong economic growth and inflation. (A smaller number of REITs are mortgage REITs; they own pools of property loans. There also are a few hybrids.) Equity REITs build or buy and own and manage properties—residential, commercial, industrial, and developing, or any combination. Because they actually own these properties, their profits go up sharply during periods of faster growth and higher inflation, when real estate gets a boost. Not only does the underlying value of their holdings go up, but they can charge higher rents, while the value of the debt they have incurred is inexorably eroding. It's a multiple win.

When economic growth slows, though, REITs are vulnerable. In particular, those that consist primarily of retail properties such as shopping malls may be hard hit. Similarly, a REIT whose properties are concentrated in a geographic region particularly affected by an economic downturn—think Silicon Valley—may get pummeled. Those most likely to suffer, though, are ones whose balance sheets are most weighted down with debt.

In selecting a particular REIT, as with any company, the first key is to look for top-notch management that has proved itself in good times and bad. A record of consistent dividend increases over the past fifteen years is one good sign. So are steadily rising profit margins.

We'd also suggest looking for REITs that are diversified

geographically as well as in the kinds of properties they own. For instance, one with a mix of residential, commercial, and industrial properties can benefit from the relative security of apartment rents and the faster growth of commercial rents. But it's a judgment call: a more narrowly focused REIT with proven management and a long record of success may be equally attractive. Other things to look for are high occupancy rates—not much below 95 percent—and debt that is no more than 50 percent of capital. A REIT with low debt can take advantage of good times to expand its holdings, and it can get by in weaker economies.

REITs have been consistently strong performers in times of high inflation and/or negative real interest rates. If, as we expect, the coming period will be characterized by one or both of those conditions, we expect REITs to strongly outperform most stock averages, though they might lag other inflation stars such as gold and energy. But one of the nicest things about investing in the real estate arena is its relative safety. It is less volatile than many other investments, and that could prove comforting in the years ahead.

Finally, REITs are particularly appropriate for investors interested in income, because they are required to pay out most of their annual profits in dividends. All in all, if they are likely to be somewhat less exciting than some of our other recommendations, they offer a solid blend of growth, inflation protection, safety, and income.

One of our favorite REITs is Apartment Investment and Management, which concentrates on buying and managing multifamily apartment complexes. It is geographically diverse, with interests in nearly every state. Another top pick is Duke Realty, which has a diversified portfolio of properties rented to a wide range of businesses, including those in manufacturing, retailing, wholesale trade, and professional services. It also owns or controls substantial acreage ready for development. In addition, it provides, on a fee basis, a variety of services to tenants at properties owned by others.

Three Special Stocks

Like a Barnum and Bailey circus where there is excitement in more than one ring at a time, even a world where a lot of the action revolves around energy shortages and rising energy prices will contain other trends with investment potential. In the years ahead one of the strongest of these will be the aging of America. As baby boomers inch into their mid- to late sixties and beyond, the median age of this country will continue to rise. In fact, virtually every other industrialized nation in the world, as well as China, is aging. This will put the spotlight on, among other things, health care. You might think that this would make it smart to invest in companies in the business of health, such as the major drug companies, and we used to think so too. In fact, our drug recommendations have been sterling performers.

There are some problems, however, with continuing to invest in pharmaceutical and other health care companies. In particular the widening U.S. deficit is creating tremendous pressure to keep health care costs under control, and this means that any health care company in any way dependent on government outlays is vulnerable to cutbacks. Those companies whose sales growth is already slowing will be the most vulnerable. While the major drug companies no doubt will experience strong unit growth in the years ahead, that is, they will sell more drugs, the pressure to keep some sort of lid on drug prices could slow their profit growth somewhat.

Fortunately, there are other companies equally leveraged to the same demographic trends but whose profits are independent of the government. An outstanding example is Weight Watchers International, which tackles head-on one of this country's most serious and visible health threats—obesity. In the past decade, obesity has increased by a third in the U.S., while the weight of the typical American has risen by more than seven pounds. Obesity is directly linked to an array of health hazards, including the diabetes epidemic, heart disease, stroke, even cancer. It is also becoming increasingly well documented that low body weight correlates strongly with longevity.

The weight loss market is nearing $40 billion and growing rapidly. By a wide margin, Weight Watchers, a worldwide franchise that offers support groups, a well-accepted weight loss system, and weight loss products, is the surest way to play the compelling need to control obesity. Originally part of Heinz, Weight Watchers was spun off by a leveraged buyout in 2001. Since then every important metric, from revenues to free cash, has grown by 20 percent or more a year. We expect earnings to grow at a 25 percent annualized rate over the next five years. As debt from the leveraged buyout is paid back, free cash, which is money available for dividends, share repurchases, or other purposes, will grow even faster.

How does Weight Watchers compare with the

franchises that make up the bulk of Warren Buffett's portfolio, such as Coca-Cola? The critical difference is growth prospects. While in terms of its P/E Weight Watchers is every bit as expensive as Buffett's holdings, it is much smaller relative to its potential market and therefore has much greater growth potential. Capitalized in 2003 at less than $5 billion, Weight Watchers fits into the category of relatively small-cap stocks while also giving you a stake in a key demographic trend.

A second small-cap stock worth singling out is Tiffany, a deep franchise in luxury goods retailing. Tiffany and its distinctive blue boxes have been a symbol of high quality both in jewelry and in other luxury goods for more than 150 years. As gold and other precious metals rise in value, the demand for jewelry is likely to increase, as buyers see it not just as a lovely thing to own but as an investment as well. In addition, as precious metals and jewelry become ever more prized, Tiffany is likely to gain an increased ability to raise prices as it wishes. Despite being a relatively small company, its franchise status reaches far beyond U.S. borders. Indeed, historically the Japanese market has been extremely profitable for the company. Given Asia's traditional love of gold, we'd be surprised if Tiffany didn't establish a beachhead in China as well. The fact that the company held its own during the slow growth period of 2000–2003 bodes well for its prospects in the period ahead, in which inflation and prices for real assets, including jewelry, are likely to be on the rise.

Our third off-the-reservation pick is a tech stock, in fact our longtime favorite tech stock, Intel, which is the only tech stock we would consider as a long-term holding. Intel has two compelling characteristics. First, it dominates a critical technology area, microprocessors, which are the brains of computers. The company's enormous financial resources and manufacturing expertise virtually ensure that it will remain dominant and indeed extend its hegemony into other tech areas such as semiconductors designed for a wireless Internet. As one indication of why it should be able to maintain its edge, just consider the fact that Intel's capital expenditures are consistently larger than its biggest competitor's revenues.

One of the most striking things about Intel is its longevity. The semiconductor industry was born in the late 1940s with the invention of the transistor. Intel's former chairman and largest individual shareholder, Gordon Moore, began his career working with one of the inventors of the transistor. Few large established companies have a lineage that traces back to the founding of their industry. That's more than just an interesting fact. For few sectors of the economy are as cyclical and turbulent as technology, and if you have proved that you can not just survive but thrive during periods when tech is severely depressed, you've earned your wings. During the tech debacle of the early 2000s Intel, like other tech companies, suffered, but unlike nearly any other it remained profitable and, remarkably enough, maintained an enormous capital expenditure program. So if you feel you want tech represented in your portfolio, we'd definitely point you toward Intel.

Even with Intel, however, we'd buy it only when it is trading under three times book value. (You can obtain this figure from a company's annual report, available online.

Or go to a public library and look it up in Value Line.)
Even the best tech stock should not be chased.

Stocks to Shun

Avoiding losers can be as important as finding winners. We could keep it simple and say avoid anything that we haven't explicitly recommended in all the preceding pages. In the turbulent times that await us, investing requires a highly disciplined, focused approach. And while diversification remains important, indeed critical, it is diversification within and among the groups that meet the criteria we have outlined above, not diversification among stocks in general. Don't grab on to a stock just because you tell yourself you can't go wrong with General Motors. You can, and many investors will.

If you think we must be mistaken, in the 1970s General Motors lost 65 percent from its high, and 3M dropped 50 percent. Procter & Gamble lost almost 50 percent in a nine-year period, and many others of this country's proudest companies were similarly pummeled—even as their earnings were rising.

In the years ahead, once again many of the best-known consumer names are likely to trail inflation. They will remain great companies, and we have every confidence that at some point in the future they will be great buys, but the coming period will not be their moment to shine.

In particular, we want to single out two groups likely to prove particularly ugly. First up are the American and European auto companies. Earlier we recommended Toyota as an attractive investment in the alternative fuels area. But its less adept and well-managed competitors will flounder. Ford, General Motors, and Daimler-Chrysler are all stocks to avoid.

Another group that gets an F is the airline stocks. The airlines have the dubious distinction of being the only industry in the history of capitalism that since its inception has failed to make a cumulative profit. This high-cost, intensely competitive industry has never managed to get its act together for more than a year or so at a time. As energy prices rise, the industry's woes will multiply. A few airlines will survive, and there may be some new entrants and some consolidations, but we think the industry is doomed to struggle fiercely to very little avail.

One final category of things to avoid should be obvious given our projections of rough sledding ahead for the general run of stocks. In the 1990s some of the most popular investments were the index funds, such as the enormously successful Vanguard 500 Index Fund. It was the perfect choice for scads of investors, who no doubt felt they had discovered the Holy Grail—a one-shot investment in a highly diversified group of safe big-cap stocks that seemed capable of returning annual gains of 20 percent or more year after year. But the fund began to turn sour in 2000 and will continue to produce unsatisfactory returns for a while to come.

Key Points:

♦ In addition to the major inflation beaters and energy plays that should form the bulk of your portfolio, a handful of other stocks are too good to pass up.

♦ Real estate investment trusts (REITs) will benefit

from inflation while offering steady income. Look for ones with a record of success in good times and bad. Small-cap stocks also have an edge in inflationary times; small-cap funds can dilute the risks of trying to pick out a few.

◆ Three other special picks are Weight Watchers, Tiffany, and for those who want some tech, Intel.

◆ Avoid most of the big blue chips; their earnings might rise, but not rapidly enough to counter falling P/Es. Also shun the airlines, American and European automakers, and stock index funds.

CHAPTER **17**

Deflation Hedges

It's paradoxical that while rising energy prices are most likely to kick off an inflationary spiral, they also could do the reverse and trigger a deflationary spiral instead. Come to think of it, though, that's really no different from a lot of things in life—a deprived childhood, for instance, might foster despair in one person and resilience in another. No one has a perfect fix on cause and effect.

Rising oil prices could be deflationary in two ways. Remember, they are like a big tax increase with no offsetting advantages. They serve as a drag on economic growth. If they rise too high too quickly, and the economy can't adjust to them and policymakers can't respond quickly enough, a severe economic contraction could ensue. That's the whole basis of our oil indicator.

And rising oil prices also could be deflationary by first bringing on the rising inflation that we foresee. As we explained earlier, because of the immense amount of consumer debt and all its implications, policymakers are walking a tightrope. If they let inflation rise unchecked, well, inflation will rise unchecked. If they try to rein inflation in by raising interest rates, they risk lowering home prices and setting off a chain of events that, again, will lead to severe economic contraction. They have very little leeway within which to act, and a miscalculation could send the economy lurching in a downward direction.

In addition, in this highly vulnerable, highly leveraged, highly interconnected world, there's no way to rule out an out-of-the-blue happening that will strike a blow at economic growth. Whether it's another terrorist assault or an epidemic of a deadly new virus or something else, the unexpected can happen.

Thus, in the years ahead, deflation, and deflationary scares, when it seems as if a steep economic downturn may be just around the corner, will be the counterpoint to rising inflation. The potential for these abrupt zigs and zags is at the heart of why the coming years will be ones of turbulence and volatility.

One result is that even when oil prices are rising at a moderate pace that is more consistent with inflation than deflation, it's still important to own some deflation hedges as protection against the unexpected. These are not just casual additional investments—they are essential protection. We're not expecting Armageddon, and deflation insurance shouldn't form the bulk of your holdings, any more than you should spend the bulk of your income on life and health insurance. But it's prudent and reassuring to own some at all times. When oil prices rise more abruptly, the idea is to add to your deflation-hedging positions and sell a hefty portion of your inflation-leveraged holdings.

Deflation in History

More than a century ago, a decade after the end of the Civil War, the U.S. experienced one of its greatest periods ever of economic prosperity. Sharply rising productivity, abundant natural resources, and ample labor combined to make 1875–91 years of extraordinary growth. GDP grew at an average rate of about 6 percent a year. And during those years prices were in a gentle downtrend, otherwise known as deflation.

So why are we talking about deflation as if it's such a terrible thing? Because the deflation we would experience in the twenty-first century would bear little resemblance to that of the late nineteenth century. Today deflation would likely set off a full-fledged, self-perpetuating depression.

The main difference between the U.S. in those days of yore and the U.S. today is debt. Today, as we outlined in chapter 5, the U.S. is awash in consumer debt. And when deflation strikes, debt becomes increasingly expensive. You have to pay back the same amount as before deflation emerged, but the dollars you're using become more and more valuable. Deflation can turn tolerable debt into debt that is crushing. And when this phenomenon strikes consumers and businesses en masse, forcing them to cut back on spending, the upshot is a self-feeding economic decline.

If you remove debt from the equation, deflation is less fearsome. In fact, it can simply reflect sharp productivity gains. And this means that even though incomes are rising slowly or not at all in nominal terms, they are rising in real terms. Under these circumstances, deflation leads to a virtuous circle in which gains in purchasing power mean greater gains in economic activity and further gains in purchasing power. This was what happened in the U.S. in the late nineteenth century.

A quick word on Japan also is in order. Japan has been locked in a deflationary economy for about a decade, and the country has struggled, veering from recession to recession. But the economy hasn't altogether imploded. In fact, many businesses in Japan have flourished, and even Japanese branches of American companies, including luxury retailer Tiffany, have done well. What has prevented a total deflationary collapse has been Japan's low level of consumer debt. While lately the Japanese have reduced their savings rate, their overall consumer indebtedness is far less than ours, which has kept a bad situation from being a lot worse.

But when we speak of deflation in the U.S. in the early twenty-first century, we're speaking of potential economic catastrophe. And if our oil indicator signals its onset, it's time to rush into deflation hedges.

Cash, Bonds, and Zeros

Historically the only investments that perform well during the kind of economy-ravaging deflation that would occur this time around are fixed income instruments such as cash and bonds—and in particular zero coupon bonds. The most analogous period is 1929–32, and as figure 17a, "Bonds in the Depression," shows, fixed income investments were the only shelter.

More recently, deflationary fears arose when oil prices surged and acted as the catalyst that punctured the tech bubble. The sharp fall in the market threatened to cause an economic meltdown. And from mid-1999 through early 2003, bonds rose 40 percent, while zero coupon bonds scored 100 percent gains.

Let's look in more detail at various deflation hedges. The first is cash, by which we mean money put into very short-term money market accounts, and preferably those guaranteed by the government or that invest only in government securities. Cash is a natural deflation hedge be-

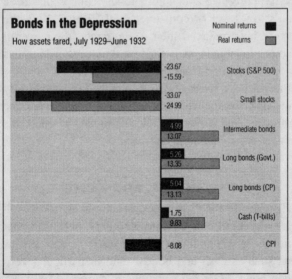

Bonds in the Depression

How assets fared, July 1929–June 1932

Nominal returns ■
Real returns ▨

	Nominal	Real	
Stocks (S&P 500)	-23.67	-15.59	
Small stocks	-33.07	-24.99	
Intermediate bonds	4.99	13.07	
Long bonds (Govt.)	5.26	13.35	
Long bonds (CP)	5.04	13.13	
Cash (T-bills)	1.75	9.83	
CPI	-8.08		

Figure 17a

cause as prices go down, the value of your money goes up—you can buy more with it. That's exactly what you want from any investment.

With bonds you get a double play. During deflationary times the capital value of bonds appreciates. That's because their payout becomes increasingly valuable as interest rates in general go down, which is what happens during deflation. In addition, the dollar income that you are getting from your bonds becomes worth more for the same reasons that cash gains in value.

Zero coupon bonds have even more potential. These are bonds that don't pay coupons at regular intervals during their life span. Rather, you buy them at a discount to par and they are guaranteed to mature at par. For instance, you could buy a zero coupon bond that is guaranteed to pay you $10,000 in fifteen years. Your purchase price, say, is $2,000. During inflationary times, this isn't so attractive, because in fifteen years $10,000 might be worth next to nothing, maybe even less than the $2,000 you put up initially. You gain little in real terms or even lose. But during deflation, they are suddenly a great deal because that guaranteed $10,000 at the end of the rainbow keeps gaining in value.

Typically during times of deflation, the gains from zero coupon bonds are 50 percent higher than the gains from regular coupon-paying bonds. The reason is that they are more leveraged to changes in interest rates. Here's why. As we noted, the value of straight bonds goes up as interest rates go down. The rise in value, though, tends simply to reflect the yield disparity; it's a straightforward adjustment. For instance, suppose a bond pays a coupon equal to $100 a year. If interest rates are 10 percent, the bond will be valued at $1,000. If interest rates fall to 9 percent, the bond will simply rise to the level where its yearly $100 coupon represents an approximately 9 percent yield. This would bring its value to around $1,111. That's a little more than an 11 percent gain in the value of the bond.

With zero coupon bonds, the same rate change will result in a larger gain in value. That's because when interest rates change, the price of the zero has to reflect all future imputed interest rate payments. The difference between the zero's current price and the price at maturity represents the value of all interest payments. When interest rates fall, the value of the bond will rise considerably more than would be the case for a bond that pays coupons. The greater the time to maturity, the greater the gain.

Why not simply put all your money into zeros? Because as we've noted, nothing in the market is ever a hundred percent certain, and if inflation picks up unexpectedly, your zeros will tank. Regular bonds will come down less, and they'll still give you income.

Individual bonds generally aren't as liquid as stocks, and sometimes you can be forced to overpay for them. Thus, when it comes to investing in any fixed income instrument as a deflation hedge, we prefer the mutual fund route. One key rule is to go for quality. Don't get tempted by high-yield bond funds, because these consist of the debt of risky companies, exactly the kind that could go broke if a depression hits. Stick to government bonds and ultra-high-quality corporate bonds.

For regular bonds, our first choice is the Fidelity Investment Grade Bond Fund (1-800-544-8888), a well-managed fund that invests in high-grade bonds. As for zero coupon bonds, we'd recommend the American Century Zero-Coupon Bond Funds (1-800-345-2021).

Key Points:

- In the coming years of economic and market volatility, deflationary scares will be the counterpoint to inflationary pressures. Investors need to hold some deflation insurance at all times. When our oil indicator flashes a negative signal, emphasize deflation plays more heavily.
- Deflation plays include T-bills, regular bonds, and zero coupon bonds. Zeros will appreciate most sharply during deflationary interludes but during inflationary stretches will offer nothing; T-bills and coupon-paying bonds provide steady income.

CHAPTER 18

Putting It All Together

Here we tie it all together for you. After all, in the preceding pages we've recommended quite an array of investments, from the most blatant inflationary plays to pure deflation hedges, from a small silver mine in Bolivia to giant Berkshire Hathaway, from a company engaged in making bombs smarter to a company dedicated to making people thinner. We think every company we've picked has extra-special credentials, and we hope we've made a compelling case for them all. But we can understand if you'd like some guidance on how to put them together.

We offer two model portfolios, one for when our oil indicator is positive, meaning inflationary forces will be prevailing, and one for when it turns negative and deflation becomes the most immediate concern. As you can see, they contain the same investments but with significantly different weightings. It's not a question of two entirely different animals but of markedly shifting your emphasis.

Actually, it takes a fair amount of courage to offer a model portfolio in a book, as opposed to in an investment letter that is updated every couple of weeks or so. When something is in a book, it is cast in stone for the ages. But we took that risk in 1999 with our last book, *Defying the Market,* and our offerings have held up remarkably well. In fact, between July 1999 and April 2003, our model portfolios from that book outperformed the market by nearly 40 percentage points. Specifically, during a period that encompassed one of the worst bear markets in history, our portfolios posted roughly 10 percent gains versus a nearly 30 percent loss in the S&P 500. Not that we're bragging, but that's a pretty decent record. Figure 18a, *"Defying the Market* Portfolio Performance," shows the results from following one of the portfolios from that book. It happens to be one that was geared toward more conservative investors with a large amount of money to invest, but the results from any of the other portfolios are comparable.

Looking at the portfolio's performance over the four years between 1999 and 2003 is revealing and offers some hints about how various kinds of investments might perform in the years ahead. One noteworthy point is that both our inflationary and deflationary hedges did well. It was the broad middle—those stocks that weren't particularly tied in to inflation or deflation—that, like the S&P 500, got clocked. As we explained in chapter 10 on the basics of inflation investing, the general run of stocks should at best perform roughly in line with inflation and at worst should sharply underperform. The fact that we mostly stayed away from that bland middle area is the reason that our portfolios performed so well.

Figure 18a: *Defying the Market* Portfolio Performance

	Symbol (Bloomberg)	Starting value (July 1999)
Deflation		
Straight bonds	VFITX	$240,000
Zero coupon bonds	BTTRX	$180,000
Berkshire Hathaway	BRK/A	$200,000
Duke Energy	DUK	$20,000
Franchises		
Coca-Cola	KO	$160,000
Gillette	G	$160,000
Disney	DIS	$80,000
Pfizer	PFE	$60,000
Merck	MRK	$60,000
Berkshire Hathaway	BRK/A	$160,000
Inflation Hedges		
Small stocks	RUV	$120,000
REITs	BBREIT	$80,000
Gold & gold stocks	XAU	$80,000
Energy service cos.	OSX	$60,000
Large energy stocks	XOI	$140,000
Food stocks	S5FDPR	$80,000
Environment		
Berkshire Hathaway	BRK/A	$80,000
Enron	ENRNQ	$20,000
Thermo Electron	TMO	$20,000
Total		$2,000,000

Gross dividends reinvested performance of recommended portfolio. S&P 500 gross dividends reinvested performance.

Recommended portfolio relative performance to S&P 500

*Returns are gross dividends reinvested performance.

% weighting	% return*	Ending value (April 2003)
12.0%	40.83%	$337,982
9.0%	97.38%	$355,278
10.0%	6.04%	$212,079
1.0%	−31.63%	$13,675
8.0%	−29.54%	$112,732
8.0%	−22.51%	$123,986
4.0%	−30.72%	$55,424
3.0%	−3.34%	$57,997
3.0%	−10.65%	$53,609
8.0%	6.04%	$169,663
6.0%	−8.93%	$109,283
4.0%	46.32%	$117,060
4.0%	15.17%	$92,138
3.0%	8.07%	$64,845
7.0%	−6.37%	$131,089
4.0%	7.60%	$86,080
4.0%	6.04%	$84,832
1.0%	−100.00%	$0
1.0%	30.38%	$26,077
100.0%		$2,203,828
	10.19%	
	−29.16%	
	39.35%	

This time around, however, one difference is that we don't necessarily expect our inflation and our deflation positions to both do well at the same time. The years between 1999 and 2003 were unusual in that policymakers were fighting the alarming economic weakness brought on by the bursting of the tech bubble. The result was a combination of deflationary forces and fears, on the one hand, and inflationary policies, on the other, a schizophrenic mix that made investors grab on to both deflation and inflation plays.

And this was true even though inflation itself remained tame (though as we've noted earlier, not quite as low as government statistics would have you believe), dampened by the effects of the tech meltdown. To explain it more precisely, the tech bubble left in its wake a massive amount of excess capacity in many industries; in addition, the huge loss in wealth from the battered stock market made consumers gun-shy about spending too freely. In this context, even the protracted period of negative real rates in combination with rising energy prices didn't boost overall inflation to much higher levels.

This helps illustrate the point that some inflation plays can do very well even if inflation itself does not accelerate. This is particularly true for precious metals such as gold and silver and for real estate and real estate investment trusts. As we explained earlier, while real assets like these tend to follow inflation, the true impetus for their gains comes from negative real interest rates.

Right now, though, we don't foresee another bubble the size of tech emerging and then crashing, which makes it less likely that we'll end up with a situation where both inflation and deflation plays will be winners. Without such a rerun of the late 1990s, deflationary plays are not likely to do nearly as well in the next few years as they did in the recent past. But we are in a turbulent "who knows what will happen" world that will be made ever more so by the worsening energy squeeze. Homes, for instance, could turn into the next bubble at some point in the future. We're not predicting this will happen, but it can't be ruled out either. Thus, deflation hedges are essential insurance, even when our oil indicator is positive. Anytime it turns negative, they should become your biggest position.

If we ever do get another massive bubble, the thing most likely to prick it will be a sharp rise in oil prices, the ultimate reality test. We have little doubt that future historians will look at the stock market and economy of the 1990s and early 2000s and proclaim that rising energy prices were behind the undoing of the tech boom. Tech stocks soared because investors bought into the belief that economic growth could accelerate indefinitely. The sharp rise in oil and natural gas prices punctured that belief, and in doing so punctured the bubble. If home prices become the next bubble, investors in homes may come to feel that with oil prices rising so rapidly, the gains in home values will be unsustainable because of the downward pressure on the economy. A groundswell of selling could feed on itself and overwhelm even very negative real interest rates.

Structuring Your Portfolio

In the years ahead, in which extremes of either inflation or deflation can't be ruled out, investors need to be alert, flexible, diversified within a carefully selected assortment of inflation and deflation plays, and leveraged to get the greatest gains from the prevailing economic winds. It's not as hard as it sounds, thanks to our oil indicator.

We'll start with flexibility. Our oil indicator is the basis for distinguishing between two types of investment environments, inflationary and deflationary. We're expecting an inflationary environment for far more of the time, but such an environment is inherently unstable, and deflationary expectations can loom up at any time.

Let's say we're in an inflationary mode and you have bought the stocks we suggest in roughly the weightings we suggest. You're keeping tabs on oil prices by checking their year-over-year changes, figures you can find in the *Economist* each week or get online or from your broker. Anytime oil prices are 80 percent higher than year-ago levels, the indicator falls into negative territory and you should dramatically lighten up on your inflation holdings and add to your deflation holdings. When the indicator shifts back into positive mode—a year-over-year change of 20 percent or less—the biggest concern once again becomes inflation. Recalibrate your holdings to reflect the change.

Why not get out of all inflation plays when the indicator is negative and all deflation plays when it's positive? The answers are implicit in what we've already discussed. We think our indicator is the best single gauge you can use to predict the market, but that doesn't mean it's perfect—no indicator is. Deflationary shocks can happen at any time, even without a warning from our indicator. And because of the effects of negative real interest rates, many of our inflation plays will do well even in a deflationary environment.

Figure 18b: Model Portfolios

Investments	Oil indicator positive	Oil indicator negative
Energy	23%	10%
ChevronTexaco		
Nabors		
PetroChina		
Schlumberger		
Energy funds		
Gold	15%	10%
Newmont Mining		
Barrick Gold		
Gold coins		
Alternative energy	20%	5%
General Electric		
Toyota		
Petro-Canada		
EnCana		
Apex		
Silver/platinum coins		
Defense	8%	5%
Northrop Grumman		
General Dynamics		
Berkshire Hathaway	15%	15%
Miscellaneous	14%	5%
Small-cap funds		
Weight Watchers		
Tiffany		
REITs		
Deflation	5%	50%
Zero coupon bond funds		
Cash		

The portfolios are diversified among the various areas covered in our investment chapters. You'll note that for many groups we've offered a representative sample of the stocks recommended in the chapters rather than including every stock mentioned; we also for the most part have stressed individual stocks rather than funds. We've picked our current favorites, the ones we think are the surest and the most leveraged to economic trends, but that doesn't mean we don't still like the others. Depending on the size of your portfolio, you can include more or all of them, and we'd have no problem if you decided to invest largely through the mutual funds we have singled out earlier. And, of course, it's important to continue to monitor your picks to make sure they stay on track. We'll be doing that for clients and in our investment letter, and you should too.

Finally, you might ask, what if we're completely wrong? What if oil prices don't rise, what if the world economy grows at a slow and steady pace or even at a rapid and steady pace, without much inflation and the various imbalances we foresee? How would our portfolios stand up then? We actually still think you'd be in pretty good shape. Our picks wouldn't be sharply leveraged to underlying economic trends, which means they would be less likely to sharply outperform, and some might not pay off for a while. But we've selected our investments very carefully and they are not lightweights. So even if we're wrong—and we don't think we are—about what lies ahead, you'll still own a collection of great companies, not a bad position to be in.

Our Future

Arthur C. Clarke is best known as a science fiction writer, in particular as the author of *2001: A Space Odyssey*. He also, however, is a well-respected scientist and philosopher who was knighted for a variety of achievements, among them the invention of the communications satellite. Now in his eighties, he has been living for some time in Sri Lanka. In the fall of 1999 the science section of the *New York Times* carried a fascinating interview with Clarke about his thoughts on the future of the universe and the human race. That interview was the catalyst for this book.

One question put to Clarke was how he felt about the recent announcement that the world's population had reached six billion. He replied: "Well, I feel rather depressed, but then there are so many times when I'm an optimist. I think we have a 51 percent chance of survival. I would say the next decade is perhaps one of the most crucial in human history, though many people have felt that way in the past. But it's real now. There are so many things coming to a head simultaneously. The population. The environment. The energy crunch. And, of course, the dangers of nuclear warfare. I am often asked to predict things, and I'm described as a prophet, but I deny that. I'm just an extrapolator. I can envision a whole spectrum of futures, very few of which are desirable."

Those words resonated with us. And as we thought about them, it seemed that of the four issues he singled out—the population explosion, the environment, energy, and nuclear warfare—in many ways energy was the most important, because it could be the key to solving the rest. With respect to population growth, if we could develop clean, renewable energies, we'd be able to desalinate the ocean and provide adequate food and water for growing populations. Moreover, as nations become more economically developed, a process that cheap energy would foster, population growth invariably slows down. Finally, once energy became abundant and cheap, our space program would have no limits. Colonizing other worlds, which doubtless seems very far-fetched and remote today, could happen—but if and only if we develop cheap and abundant energy.

As far as the environment goes, if we stopped burning coal and oil, we'd instantly end the major source of environmental degradation. And a world with limitless energy is also a world far less threatened by nuclear warfare, for several reasons. First, poverty would diminish, and there'd be far less reason for poorer nations or people to seek to take up arms against wealthier nations. Also, once we're no longer economically hostage to unstable and hostile oil-producing nations in the Middle East and elsewhere, we'll be far more able to root out any terrorism that remains in those areas. Finally, it would become feasible to dismantle all nuclear plants, eliminating any further production of weapons-grade plutonium.

Even before reading the interview with Clarke we had begun to focus on energy. In our last book, *Defying the Market,* we began to lay out the case that oil supplies were slated to lag rising demand, which is why we strongly recommended the then bottom-dwelling energy stocks. Since then, it has seemed increasingly clear not just that we were right but also that the energy crunch would start to intensify even more quickly than we had thought. That realization, fanned by Clarke's comments, led us to really zero in on the whole area of energy and energy alternatives.

In the preceding chapters we mostly took our views about energy and distilled them through an investment perspective. We looked at the effect of rising oil prices and energy shortages on the economy and through the economy on the financial markets. We think we see what's coming, and we want to help investors deal with it in the best way possible.

But though it may be heresy to say this in an investment book, there's more to life than investing. Even if we all could make a bundle by following the advice in this book and capitalizing on trends in energy prices, it's ultimately not going to do us much good if this country botches the task of developing energy alternatives. Everyone's long-term welfare depends on how intelligently, nonpartisanly, and quickly we address our energy problems. And the record up to now has been, frankly, abysmal.

Real Men Burn Oil

One problem has been that for a long time the whole issue of alternative energies has been seen largely in the context of an overall political gestalt. If you're the kind of person who cares about alternative energies, you probably also are a vegetarian or at the very least are into organic foods, and chances are you're some kind of artist or maybe a teacher or social worker. If you're made of tougher stuff, a construction worker or a corporate executive, you know that it's oil that makes the world go round.

This dichotomy was perhaps best symbolized by the actions of two presidents. Jimmy Carter, he of the professorial-like cardigan sweaters, installed solar panels in the White House. Ronald Reagan, the cowboy president, ripped them out.

Today, however, the issue of energy alternatives is catching up with us. We're running out of time, and as oil production peaks, we no longer will have the luxury of treating alternative fuels as a matter of politics or style or as something we can leave for a later generation. Instead, it will become a central issue for us all. To put it even more strongly, the survival of our capitalist system will depend on our ability to resolve it. If that seems too extreme a statement, it's not. Our system can't survive without affordable energy, and once oil doesn't fit that bill, we'll need to find something else that does. We'll all have to come together on this.

Rising oil prices will force the issue. But even then we can't take for granted that we will get on the right path. We can't rely on good old-fashioned capitalism and the profit motive to produce solutions, to generate the best technology or create the necessary infrastructure. Private enterprise will play a big role, but it can't do it all on its own.

America's signal accomplishments during the twentieth century—whether winning wars, building the interstate highway system, or landing a man on the moon—required efforts well beyond the reach of private enterprise. They required intense and massive cooperation between the private and public sectors. Developing alternative energies will be no different.

This will be an enormous challenge and also an immense opportunity. Just as the space program and the highway program created many new industries in this country and helped promote economic growth, so will the development of alternative energies. And whether we like to admit it or not, this is a country in dire need of new industries. As an ever greater portion of our manufacturing base has been transferred to low-labor-cost countries such as China, our leadership rests almost entirely on our superiority in defense. Is this a large enough peg? Maybe over the short term. But without energy, even the strongest defense apparatus will grind to a halt.

If the U.S. can establish a leadership position in the development of alternative energies, we'll not only have successfully met the tremendous challenge of declining oil supplies. We'll have built a vital new industry based on vital new technologies that will spur economic growth and ensure our continued hegemony in the world community.

Moving On

What's required? Making a full-fledged commitment to wind energy by helping to fund a nationwide network of turbines, while funding long-range research into improved solar cell technology and hydrogen storage. Having a clear goal, as we did in putting a man on the moon. And launching this effort now. It's already late in the day. The sooner we make a real commitment, the better, because there will be no deus ex machina to bail us out. Token funding of alternative energies won't be enough. Our whole national mindset has to change.

It would have been nice, of course, if the big push to develop alternative energies had been made before we reached this pass. Even before the first Arab oil embargo it was clear to some people that it made no sense to rely so heavily on fossil fuels. Buckminster Fuller was an early visionary in this area, as in many others. In 1969 he wrote in his book *Utopia or Oblivion*: "There are gargantuan energy-income sources available which do not stay the processes of nature's own conservation of energy within the earth crust 'against a rainy day.' These are in water, tidal, wind, and desert-impinging sun radiation power. The exploiters of fossil fuels, coal and oil, say it costs less to produce and burn the savings account. This is analogous to saying it takes less effort to rob a bank than to do the work which the money deposited in the bank represents. The question is cost to whom? To our great-great-grandchildren, who will have no fossil fuels to turn the machines? I find that the ignorant acceptance by world society's presently deputized leaders of the momentarily expedient and the lack of constructive, long-distance thinking—let alone comprehensive thinking— . . . render dubious the case for humanity's earthian future."

There are many reasons for our failure to move ahead

on the alternative energy front sooner. The factors, for sure, would include the intricate realities of Middle East politics and diplomacy and the clout of oil companies. But in the final analysis it likely comes down to the inherent short-sightedness of the human race, our disinclination to endure immediate pain in the hope of long-term gain. As Aesop knew, when it comes to planning ahead, we're almost all grasshoppers.

For a long time, the simple fact was that oil and other fossil fuels were a lot cheaper than any prospective alternatives. For alternative fuels to have made it onto the playing field, the government would have had to step in and take action, such as steeply raising the price of oil through taxes. One way or another, such steps would have made life tremendously more expensive for the entire voting public, not an intrinsically appealing idea to your typical politician. Even if, say, an enlightened president was interested in doing this, getting a majority of Congress to go along would not have been a good bet.

But if it's frustrating that up to now we've put off facing this issue in any meaningful way, we aren't abandoning hope. The following quote may offer some reassurance: "A country was running out of an important natural resource. . . . The rapid depletion, brought about by the huge increase in demand, caused the price of the resource to rise rapidly. In response to the resource shortage and consequent high prices, people undertook technological change and found a substitute that turned out to be preferred to the original resource at its new, higher price."

These words were written to describe an earlier transition from one energy source to another—England's shift in the sixteenth century from firewood to coal.

Written in 1984 by Charles Maurice and Charles Smithson in their book *The Doomsday Myth,* they are a valuable reminder that the world has had to cope, and has done so successfully, with major energy crises in the past.

And they point the way to what will finally change things—rising oil prices. As oil prices continue to go higher, the need to move beyond oil will finally get everyone's full attention. There have been oil scares in the recent past. But as we hope we've shown, this one will be different—this one isn't just going to go away. As Clarke said in the passage quoted above, this time it's real. We are facing a momentous transition, one that requires real leadership and massive effort and funding. This transitional period will be marked by both turbulence and opportunity. Most important, unless we truly blow things, it ultimately should result in a far better world, one where growth is based on renewable, nonpolluting energies that can create happier lives for everyone on this planet. That truly will be a revolution worth participating in.

One final note: in our last book, *Defying the Market,* we argued that despite all the hoopla over such marvels as the Internet and the human genome project, technological progress was slowing and true scientific breakthroughs were becoming scarcer. So is it inconsistent to now insist that we can employ the necessary technologies to overcome our energy problems? Not at all. For the solutions to dwindling oil and gas at least over the next couple of decades don't lie in new technological breakthroughs but rather in utilizing technologies that have been around for a while—in the case of clean coal, for more than thirty years, and in the case of wind, for many centuries.

Appendix
List of Recommendations

Stocks

Apartment Investment and Management (AIV) NYSE
Apex Silver (SIL) NYSE
Barrick Gold (ABX) NYSE
Berkshire Hathaway (Class B—BRK) NYSE
Boeing (BA) NYSE
CACI International (CAI) NYSE
ChevronTexaco (CVX) NYSE
Devon Energy (DVN) AMEX
Duke Realty (DRE) NYSE
EnCana (ECA) NYSE
Exelon (EXC) NYSE
General Dynamics (GD) NYSE
General Electric (GE) NYSE
Headwaters (HDWR) NASDAQ
Impala Platinum (IMPUY) NYSE
Intel (INTC) NASD
Lockheed Martin (LMT) NYSE

Nabors (NE) NYSE
Newmont Mining (NEM) NYSE
Noble (NE) NYSE
Northrop Grumman (NOC) NYSE
Petro-Canada (PCZ) NYSE
PetroChina (PRT) NYSE
Raytheon (RTN) NYSE
Schlumberger (SLB) NYSE
Tiffany (TIF) NYSE
Toyota Motor (TM) NYSE
Weight Watchers International (WTW) NYSE

Funds

American Century Global Gold Fund (BGEIX)
American Century Zero-Coupon Bond Funds:
       Target Maturities Trust: 2005 (BTFIX)
       Target Maturities Trust: 2010 (BTTNX)
       Target Maturities Trust: 2015 (BTFTX)
       Target Maturities Trust: 2020 (BTTTX)
       Target Maturities Trust: 2025 (BTTRX)
       Target Maturities Trust: 2030 (ACTAX)
       1-800-345-2021;
       www.americancentury.com
Excelsior Energy and Natural Resources Fund
   (UMESX)
       1-800-446-1012;
       www.excelsiorfunds.com
Fidelity Investment Grade Bond Fund (FBNDX)
Fidelity Low-Priced Stock Fund (FLPSX)
       1-800-343-3548; www.fidelity.com

Gabelli Gold Fund (GOLDX)
        1-800-422-3554; www.gabelli.com
ICON Energy Fund (ICENX)
        1-888-389-4266; www.iconfunds.com
Royce Total Return Fund (RYTRX)
        1-800-221-4268; www.roycefunds.com
Third Avenue Small-Cap Value Fund (TASCX)
        1-800-880-8442; www.thirdave.com
Tocqueville Gold Fund (TGLDX)
        1-212-698-0800; www.tocqueville.com
Vanguard Energy Fund (VGENX)
Vanguard Precious Metals Fund
        1-877-662-7447; www.vanguard.com

To buy silver and platinum directly, use the Web site www.kitco.com.

Index

About the Authors

STEPHEN LEEB is the president of Leeb Capital Management, the editor of *The Complete Investor*, and the author of four previous books. His wife and longtime collaborator, DONNA LEEB, has a diversified background in business writing and holds a master's degree in journalism from Columbia University.